PATH PLAYER GAMES

Springer Optimization and Its Applications

VOLUME 24

Managing Editor
Panos M. Pardalos (University of Florida)

Editor—Combinatorial Optimization
Ding-Zhu Du (University of Texas at Dallas)

Advisory Board
J. Birge (University of Chicago)
C.A. Floudas (Princeton University)
F. Giannessi (University of Pisa)
H.D. Sherali (Virginia Polytechnic and State University)
T. Terlaky (McMaster University)
Y. Ye (Stanford University)

Aims and Scope
Optimization has been expanding in all directions at an astonishing rate during the last few decades. New algorithmic and theoretical techniques have been developed, the diffusion into other disciplines has proceeded at a rapid pace, and our knowledge of all aspects of the field has grown even more profound. At the same time, one of the most striking trends in optimization is the constantly increasing emphasis on the interdisciplinary nature of the field. Optimization has been a basic tool in all areas of applied mathematics, engineering, medicine, economics and other sciences.

The Springer Series in Optimization and Its Applications publishes undergraduate and graduate textbooks, monographs and state-of-the-art expository works that focus on algorithms for solving optimization problems and also study applications involving such problems. Some of the topics covered include nonlinear optimization (convex and nonconvex), network flow problems, stochastic optimization, optimal control, discrete optimization, multiobjective programming, description of software packages, approximation techniques and heuristic approaches.

PATH PLAYER GAMES

Analysis and Applications

By

SILVIA SCHWARZE
Department of Business and Economics
University of Hamburg, Germany

 Springer

Silvia Schwarze
Department of Business and Economics
University of Hamburg, Germany
schwarze@econ.uni-hamburg.de

ISSN: 1931-6828
ISBN: 978-1-4419-4606-5   e-ISBN: 978-0-387-77928-7
DOI 10.1007/978-0-387-77928-7

Mathematics Subject Classification (2000): 90-02, 90B10, 90B20, 91-02, 91A10, 91A43

Cover illustration: Cover photo taken by Torsten W. Schneider

Printed on acid-free paper

springer.com

To Torsten,
I am glad that we play on the
same path.

Contents

Symbols and Abbreviations

General Symbols

$\mathbf{0}_n$ .... n-dimensional zero vector
$\mathbf{1}_n$ .... n-dimensional one vector
$\mathbb{I}_n$ .... $n \times n$ identity matrix
$\mathbb{N}_0$ .. natural numbers including zero
$\mathbb{R}_+$ ...... nonnegative real numbers
$\mathbb{R}^n$ .. n-dimensional Euclidian space
$\mathbb{R}_+^n$ ..... nonnegative n-dimensional
Euclidian space
$sgn(a)$ .... signum of a

Symbols: Path Player Game

$b(f)$ ..... vector of benefit functions
$b_P(f)$ .... benefit of player P
$\tilde{b}_P(f)$ .... one-dimensional benefit of
player P
$c_e(f_e)$ .... cost on edge e
$c_P(f)$ .... cost of path P
$\tilde{c}_P(f)$ .... one-dimensional cost of P
$d_P(f_{-P})$ .... decision limit
e .... edge
e_m .... class of edges shared by
player set m
E .... set of edges e
E_P^{com} ... set of commonly used edges
E_P^{exc} ... set of exclusively used edges
f .... network flow
f_e .... flow on edge e
f_P .... flow on path P

f_{-P} ...... network flow excluding P
$f_P^{\text{max}}(f_{-P})$ ... best reaction set of P
$\mathbb{F}$ .... set of feasible flows
Γ .... game instance
G .... game network
$G(n)$. standard network of n players
$I(\varphi)$ .... cost of sequence φ
κ_P ..... security payment of path P
$l(\varphi)$ .... length of sequence φ
M ... real number, sufficiently large
$ND(\Gamma)$ ... set of nondominated flows
$NE(\Gamma)$ .... set of equilibria
ω_P .... security limit of path P
$\Pi(f)$ .... potential function
φ .... strategy sequence
P .... path
$\mathcal{P}$ .... set of paths P
$\mathbb{P}(n)$ ..... power set of set of players
r .... flow rate
s .... source vertice
t .... sink vertice
v .... vertice
v_i .... ith unit vector
V .... set of vertices v

Symbols: Game on Polyhedra

A .... coefficient matrix
b .... coefficient vector
$c_i(x)$ .... cost of player i

h_i ....strategy set of player i

$H(A, b)$ ....hypercuboid

i ....player

n .... number of players

$S(A, b)$ ....polyhedron of feasible solutions

$S_i(x_{-i})$ feasible strategies of player i

x .... solution

x_i .... strategy of player i

x_{-i} ....solution excluding x_i

$z_m(x_m)$ ......cost function of set m

Symbols: Line Planning Game

$b_P(f)$ .... payoff of line P

$c_P(f)$ .... cost of line P

$d_P^1(f_{-P})$ ....lower decision limit

$d_P^2(f_{-P})$ ....upper decision limit

f .... network frequency

f_e ....frequency on edge e

f_P .... frequency on path P

f_P^{br} ....best reaction set of P

$f^{\min}$ .... minimal frequency

$f_q^{\min}$ ..minimal frequency of $\{s_q, t_q\}$

$f_e^{\max}$ ..maximal frequency on edge e

$\mathbb{F}^{LPG}$ .....set of feasible frequencies

$\mathbb{F}^{ILPG}$ ....set of feasible integer frequencies

N ....real number, sufficiently large

P ....line

Q ....number of OD pairs

$\mathcal{P}$ ....line pool

$\mathcal{P}_q$ ....line pool of $\{s_q, t_q\}$

s_q ....qth origin

t_q .... qth destination

Abbreviations

AFIP ....approximate finite improvement property

FBRP ....finite best-reply property

FIP ....finite improvement propery

GNE .... generalized equilibria

ILPG ... integer line planning game

LPG ....line planning game

NCS ..... noncompensative-security

OD .... origin–destination

PPG .... path player game

QVI ...quasi-variational inequalities

1

Introduction

1.1 Network Games

Various types of games on networks have been studied in recent years. The variety of models is huge, reaching from basic design of a network to routing network flow or allocating resources in existing networks. For instance in routing games, flow has to be transported from origin to destination nodes. It is assumed that the flow itself consists of atomic [CCSM06, BMI06, Rou05b] or nonatomic [Rou05a, KP99] players that independently choose their way to the destination. The objective is to determine equilibria under the assumption that the network is congested, i.e., the cost for using edges is growing with the load on an edge. An important question is the loss of efficiency caused by the selfishly acting players compared to a centrally controlled scenario. Related are the Stackelberg network games (or: Stackelberg routing games) [Swa07, KS06, BS02, Rou04, KLO97], where central control is partially admitted. One or several players have a leading function and are able to influence the price, e.g., by fixing parts of the flow. The leaders maximize their own payoff taking into account the noncooperative behavior of the remaining players, the followers.

Not only routing problems are modeled as games on networks. In load-balancing games [GC05, STZ04, CKV02], load is to be assigned to resources such as jobs to machines, e.g., in server farms. The network pricing game [HTW05] combines the view of a network provider and a network user. Network providers compete for customers by offering different prices and services. In facility location games [Vet02, GS04, CCLE$^+$06, Mal07], facilities like warehouses or public utilities have to be located and allocated to the demand points. Related are the service provider games where a provider offers service to possible customers [DGK$^+$05, BCKV06]. Minimum cost spanning tree games [Bir76, DH81] and Steiner tree games [SK95, KLS05] are cooperative models that describe situations where users are connected to a common supplier via network. The goal is to find a sta-

S. Schwarze, *Path Player Games*, DOI 10.1007/978-0-387-77928-7_1,
© Springer Science+Business Media, LLC 2009

ble allocation of routing costs to the users. Finally, network design games [CRV08, AK07, CR06, FLM+03, ADK+04] and network formation games [EBKS07, GVR05, Jac05, BLPGVR04] describe the generation of networks.

Studying network games is of high interest as applications are found in various fields. For instance, telecommunication systems are a very active research field with respect to network games. Many questions of telecommunications are modeled as game situations, for instance routing problems, load balancing, resource allocation, or network security are topics of ongoing research. See Altman et al. [ABEA+06] for a recent and extensive overview concerning network games in telecommunications. Another related area where network games are applied successfully is the computer sciences. Routing, traffic allocation, load balancing, and pricing issues are becoming important everytime computers are connected through networks and when services have to be provided [Pap01, CKV02]. Agent-based models incorporate game theoretic principles like the selfish behavior of the agents [AK05, Nis99].

Naturally, network games do apply for all kind of optimization problems which can be modeled as networks. We have already mentioned facility location problems above. Moreover, questions arising in transportation networks that are modeled by network games. For instance, an often cited example for the routing game is private automotive traffic during rush hours. Furthermore, line planning in public railway systems is discussed in [SS06b] and in Chapter 4. In addition, see [JMSSM05, CLPU04, CC01] where routing and passenger assignment in transportation systems are studied. As social relations may be modeled by networks, they also serve as applications for network games; see, [GVR05, GGAM+03, GO82] for examples.

The future prospects of network games are promising. Apart from the highly active ongoing research, there are fields which might be successfully captured by networking games. For example, the investigation of network properties motivated by physical or biological applications will be a promising new area of research. The concepts of centrality, density, and connectivity of networks are worthwhile to be studied in the scope of network games; see [JS08, BE05] for an introduction into the matter. Moreover the dynamic character of evacuation problems (see [BS06, HT01, SS02]) makes them a possible application for network games, where the uncontrollable behavior of the evacuees matches the independent and selfish acting players.

1.2 The Scope of This Book

In this book, we study a new type of routing game. Usually in routing games, the problem of sending flow in a network is considered from the point of view of the flow itself, assuming that the flow can choose a path from the origin to the destination. Another interesting aspect, which has not been considered yet, is the behavior of the path owners, when they are allowed to choose the amount of flow that will be sent along the paths. This new constellation models systems where paths are owned by decision makers; the decision makers offer their paths for use by the flow. Equilibria in this model describe a stable market situation among path owners. Thus, the existence, characterization, and computation of equilibria is an important research topic and is investigated in this text. An application to a particular problem of public transport optimization is considered. Moreover, a network-independent generalization of the model is proposed.

Path player games are a new type of network games that analyze competitive situations in networks from the path owner's point of view. In many situations, networks are shared by several owners, such as service providers in information networks or suppliers in energy networks. Usually, these owners are in a competitive situation, as they want the customers to pay for the offered services (energy, bandwidth). Path player games model this competitive situation by considering the paths in a network as players in an infinite noncooperative n-person game. The strategy of each player, which is privately kept, is to choose a nonnegative amount of flow to be routed along his own path. A maximal flow rate is defined to limit the amount of flow in the network. The benefit to each player incorporates a cost function depending on the complete flow in the network. Furthermore, a penalty for infeasible flow (i.e., a flow that exceeds the flow rate of the network) is considered in the benefit. Finally, a security payment offers a benefit, which can be obtained by a player when the flow is feasible.

In this concept of an infinite noncooperative game with noncontinuous benefit functions, we prove the existence of equilibria in pure strategies. Furthermore, we analyze characterizations of equilibria for special instances of the game. In fact, path player games provide multiple equilibria in many cases. Hence, in a second approach we investigate dominance among equilibria and among flows in general. We show that there are classes of path player games where each nondominated flow is an equilibrium, and also the existence of cases where each equilibrium is nondominated. Furthermore, the equality of the set of equilibria and nondominated flows holds for another class of path player games. On the other hand, we can find cases where the set of equilibria and nondominated flows have no intersection point, a situation similar to the Prisoner's Dilemma.

Path player games turn out to be interesting also from another theoretical aspect. We prove that the class of path player games is a new class of exact potential games; that is, an exact potential function exists. Potential functions model the increase or decrease of benefit a player is experiencing while changing her own strategy. We exploit this result to develop algorithmic approaches for the computation of equilibria. One approach requires the solution of an optimization problem, and a second one uses greedy steps to determine equilibria.

After obtaining these results, we extend the concept of path player games to a more general setting, the games on polyhedra. These games are an instance of generalized equilibrium games, that is, games where the strategy set of a player is dependent on the current strategies of the competitors. Games on polyhedra cover a large set of continuous games and the results in this field are not only of interest for path player games but also for generalized games. We present instances of games on polyhedra where the set of equilibria is equivalent to the set of optimal solutions of a linear program. Furthermore, there are instances of games on polyhedra that are exact potential games, which allows us to take advantage of the properties of potential functions as we have done for path player games. To get rid of the dependencies among strategy sets, we transform the games on polyhedra to games on hypercuboids. Hereby, we use the smallest hypercuboid that contains the considered polyhedron and introduce a penalty for solutions outside the polyhedron. These games are no longer generalized equilibrium games. We investigate the relation of equilibria in games on polyhedra and games on hypercuboids.

We apply the theoretical results from path player games and games on polyhedra to the line planning problem, a problem widely studied in transport optimization. In this problem, lines have to be chosen from a given line pool and frequencies have to be determined such that the customers' demand is satisfied. As a new approach to tackle line planning problems, we use the concept of path players and let the lines be owned by players who want to minimize the delay on the line. It turns out that line planning games are a generalization of path player games, and a special case of games on polyhedra. We exploit both facts and use solution strategies developed in previous chapters to compute equilibria in line planning games. Finally, we present a numerical example, based on the German railway system. A short summary and an outlook on future research topics in this area concludes this work.

A recommendation concerning the sequence of reading this text is given in Figure 1.1.

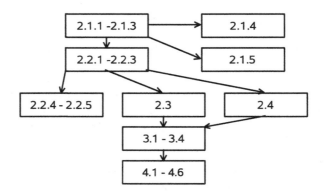

Fig. 1.1. Recommended sequences of reading.

1.3 Acknowledgments

This book is a result of joint work with Professors Anita Schöbel and Justo Puerto. Without their help, this book would not have been written and I thank them sincerely for their support. The idea of path player games came up during a stay in Sevilla, while I visited Justo. His ideas have influenced my work very strongly and I thank him for the helpful discussions and suggestions. Moreover, I am indebted to Anita. Her many good ideas and her helpful advice resulted in a very productive collaboration, which I enjoyed very much. It has been her motivation and energy that were a great help for finishing this book. Some parts of this material have been published as joint papers with Anita and Justo. This concerns most of the results of Sections 2.1 and 2.2; the introduction and investigation of equilibria of path player games are contained in [PSS08]. Parts of Section 2.3 dealing with dominance in path player games are contained in [SS06a]. Furthermore, [PSS06] presents results of Section 2.4 about potential functions. Finally, the application to line planning problems (see Chapter 4) is considered in [SS06b]. One more paper is in progress: concerning games on polyhedra, Chapter 3 is covered by [PSS].

Furthermore, there is a number of people who kindly have contributed to this book. First I want to thank Professor Horst W. Hamacher; it has been his support that enabled me to start these studies and with his help and advice I had the chance to learn a lot. I thank Horst for his enduring support that lasted longer than my time in his working group. I had two careful proofreaders for this book and I thank Dr. Alexander Lavrov and Torsten Schneider for their many helpful comments on content and language. Moreover, I thank Professor Jochen Werner for the hints on improving this text. For the numerical studies I used data from the Deutsche Bahn AG and so I thank in particular Dr. Frank Geraets for providing this material.

Finally, I send many thanks to all people that have supported and encouraged me in academic and nonacademic matters. In particular my gratitude goes to my colleagues in Kaiserslautern and Göttingen—thank you for the good times, and to my family for their patience and love.

2

The Path Player Game

In this chapter, path player games are introduced and analyzed. A rough description of path player games is the following. Given a network with a set $\mathcal{P}$ of specified paths (one for each player) and an overall flow rate r, find flows for each path not exceeding the flow rate and maximizing the benefit of the players.

2.1 The Model

2.1.1 Introduction

In this chapter, we introduce path player games, which are a new approach to model a routing problem from the viewpoint of the network operators. We consider a network that is shared by competing resource owners that represent the players in the game. These players act independently and selfishly. We assume that each player owns exactly one path, that is, a sequence of edges in the network. Each resource provider offers an amount of flow that he allows to be routed along his path; we assume a sufficiently high demand for this resource. The player receives income (or benefit) from the flow that is using his path. The benefit depends on the flow and it is not necessarily strictly increasing. Thus, situations may appear where the resource providers are not necessarily interested in routing as much flow as possible. Decreasing parts of the benefit function may be justified, for example, by increasing operating costs due to overtime, additional maintenance, and other expenses caused by handling too much flow. One interesting component of the game is that the paths may own edges shared with other paths, which can have positive or negative effects for the players. It may happen that players are forced into situations where too much flow decreases their benefit, or the other way round that the flow provided by other paths increases their income. We assume that the path owners get punished if the network is overloaded, that is, if the sum of the offered flow is higher than a given flow rate r representing the network

S. Schwarze, *Path Player Games*, DOI 10.1007/978-0-387-77928-7_2,
© Springer Science+Business Media, LLC 2009

capacity bound. In practice, this bound may arise from public regulations such as a limitation of traffic for ecological reasons, but also from capacity restrictions in the network. With this punishment, the players are forced to satisfy the feasibility constraint that requires the complete flow in the network to be less than or equal to the flow rate. If some players are too greedy, the whole group of players is punished. Nevertheless, cooperation is not allowed in path player games.

As the path player game models situations in which several providers of a commodity share a network, its applications can be found in public transport, telecommunication, or information networks. In Chapter 4, an application to the line planning problem is investigated.

The path player game is a new type of network game. We now review related literature in the field. First, there is the *routing game*, which is analyzed from the perspective of the flows. A routing game is played, like the path player game, on a *congested network*, that is, the cost functions assigned to the edges depend on the load. The flow is assumed to consist of a finite or infinite number of players. Each of the players chooses a path from source to sink that minimizes the cost of traveling along that path. This model can be seen as a counterpart to the path player game, as it represents the position of the travelers, that is, of the flow. Note that the cost functions in routing games are sometimes interpreted as latencies. An increasing latency prevents too much flow going over the edge. Thus, in some sense it acts as a substitute for the capacity constraint, which is missing in the basic version of routing games. The interpretation of cost functions in path player games is different, as we assume that the cost paid by the flow is meant as income for the path owners.

The routing game has been analyzed in several papers. A commonly discussed issue is the relation of the cost of equilibria in a routing game and the cost of a system optimum. For this relation, the notation *coordination ratio* is introduced by Koutsoupias and Papadimitriou in [KP99]. Here, a network consisting of m parallel edges is studied, which is called, according to the authors, the KP-model. The KP-model has later been studied in several publications, such as [FKK$^+$02, LMMR04, FV05]. The coordination ratio is also known as the *price of anarchy* in other references; see for example, [RT02, Rou03, Rou04] or [Rou05a], this latter containing the first references. In this material, the Wardrop model (see [War52]) is studied on networks with an infinite amount of players. The central question here and in other publications (see, e.g., [CV07]) is how to bound the coordination ratio. In [CSS04b] this analysis is done for capacitated networks. Also, some research has been done to cope with the loss of efficiency experienced by the selfish behavior of the players. One approach is to introduce taxes to regulate the traffic (see, e.g., [CDR03, Fle04, HTW05]) other attempts analyze the design of the underlying network (see, e.g., [FLM$^+$03, GKR03, ADK$^+$04]). Modifications in terms of the cost functions are considered, e.g., in [Per04]. In [CSS04a], the routing problem is considered for a different objective function, namely to minimize the maximal latency. In [KK03], side constraints such as capacity bounds are added to the routing game model. In [FV04], the authors investigate a dynamic version of the routing game, based on evolutionary processes. Routing games with multicriteria objective functions are considered, for example, in [CGY99, Nag00]. In [San01] it is shown that the Wardrop model for an infinite number of players is a potential game (see Section 2.4 for an introduction to potential functions), which is also considered by [CR06].

In terms of application of routing games, a survey in the field of telecommunications is presented in [ABEA$^+$06]. Applications to information networks are given in, for example, [CKV02, AK05, AEAP02]. In [HSK02], the authors consider in an experimental setting how forecasts of traffic jams influence the behavior of selfish-acting road users.

In contrast to routing games, the strategies of the players in a path player game can be taken as offering bandwidth to the flow. In fact, our model is related to *bandwidth allocation games*, as described, for example, in [Kel97, JT04]. In bandwidth allocation games, capacitated edges are used by several players. The players send bids to a central manager; subsequently the manager determines the prices of the edges and answers with an allocation of bandwidth that is proportional to the bids. Moreover, he cares for satisfying the capacity constraints. Each user has her own utility function that determines her payoff depending on the price and the bid. These types of games distinguish between *price-taking users* and *price-anticipating users*. The price-taking users just accept the price given by the manager, whereas the price-anticipating users take into account the reasoning of the manager and adjust their bids. Only the second approach represents a game. Contrary to this model, our model considers no capacities on the edges, although the flow rate r corresponds to a capacity in a single-edge bandwidth allocation game. Also the "bids" in the path player game (e.g., the strategies) are not answered by a manager, but are directly accepted. So, the path player game is a simpler approach which enables us to get further results. In the path player game we allow general continuous and nonnegative cost functions, whereas in bandwidth allocation games strictly increasing, continuously differentiable, and concave functions (so-called *elastic traffic*) are required. Furthermore, in bandwidth allocation games the existence of equilibria cannot be guaranteed. In the path player game we are able to prove their existence for the case of continuous cost functions.

Another model describing the behavior of path owners is the *path auction* [Yan07, NR01, NIS05, ESS04, AT02]. Here, each edge is owned by one player. A central manager has the task to buy a shortest path, leading from s to t, from the edge owners. The edge owners know the real price of their edges, but they are allowed to report a wrong price if they benefit from lying. The question is how to develop a payment mechanism such that it is in every edge owner's interest to tell the truth. Such a mechanism is called *truth telling*. This model is in a sense related to ours: assume our network consists of parallel edges from s to t; our path owners would be edge owners as well. Nevertheless,

in the path player game we are analyzing the game aspect in an earlier stage
and as a consequence we are able to obtain further results.

Parts of the results of Sections 2.1 and 2.2 have been published in
[PSS08].

2.1.2 Notation

We consider a directed network $G = (V, E)$ with finite sets of vertices
$v \in V$ and edges $e \in E$. Let the edges be given as $e = (v'_e, v''_e) \in E$, with
$v'_e, v''_e \in V$. A path P from a vertex v to a vertex $\bar{v}$ is a finite sequence of
edges: $P = (e_1, \ldots, e_K)$ such that $v'_{e_1} = v$, $v''_{e_K} = \bar{v}$ and $v''_{e_k} = v'_{e_{k+1}}$ holds for
$k = 1, \ldots, K - 1$, and such that each edge is contained in P not more than
once: $e_m \neq e_\ell \; \forall \; e_m, e_\ell \in P, m \neq \ell$. By $\mathcal{P}$ we denote the set of all paths P in G
from the single source s to the single sink t; that is, the set $\mathcal{P}$ is given by the
structure of the network G. For real-world problems, the number of players
may become very large in most cases. Hence, a variation of the game is given
by considering not all paths in a network but a subset $\bar{\mathcal{P}} \subset \mathcal{P}$. Depending on
the application, uninteresting paths could be neglected and as a result, the
problem size would decrease. Each edge e is associated with a cost function
$c_e(\cdot)$ that depends on the load on e. The cost function represents the income
of the edge owners, that is, of the paths that contain that edge. We assume
the cost functions to be continuous and nonnegative for nonnegative load:
$c_e(x) \geq 0$ for $x \geq 0$. If an edge belongs to more than one owner, we assume
that the fee is shared equally among the owners. That means, in order to cover
the payments, the flow that uses the edge has to pay a fee $K_e(f_e) = \nu_e c_e(f_e)$,
where ν_e is the number of paths that share e. It is possible to generalize this
model by allowing the owners to share the fee in an arbitrary way by introducing the share $s_{e,P}$ that path P owns of edge e. This issue is considered for
future research; see Chapter 5.

The flow is represented by a function $f : \mathcal{P} \to \mathbb{R}_+$; that is, the flow on a
path P is given by f_P.

Definition 2.1. *The flow on an edge $e \in E$ is given by the sum of the flows
on the paths containing e:*

$$f_e = \sum_{P: e \in P} f_P.$$

*The cost on a path P is given by the sum of the costs of the edges belonging
to that path:*

$$c_P(f) = \sum_{e \in P} c_e(f_e).$$

In the network, the sum of the flows is bounded by the *flow rate* $r \geq 0$
that can be interpreted as a network capacity. The flow routed from source s
to sink t shall not exceed this flow rate.

Definition 2.2. *A flow* f *is called* feasible *for a flow rate* r *if*

$$\sum_{P \in \mathcal{P}} f_P \leq r$$

holds, and infeasible *otherwise.*

It is not necessary for a feasible flow to cover the flow rate completely,[1] which also makes sense in an economic context, where the resource providers would only satisfy the complete demand if this maximizes their income, but not if the income decreases, for instance, due to overtime or additional maintenance of the resources.

2.1.3 The Rules of the Game

The paths $P \in \mathcal{P}$ in the network G represent the players[2] of the game. A finite number of $|\mathcal{P}|$ players compete with each other. Each player proposes an amount of flow f_P, his strategy that he wants to be routed along his path. Under the assumption of sufficient demand, the player implements the proposed flow. This is a considerable difference from bidding games, such as bandwidth allocation or path auction games, where the bidders receive some share that is determined by a central instance. The number of strategies is infinite as a player is allowed to choose any nonnegative real flow f_P and hence we consider an infinite game. The payoff in path player games is given by the benefit function, which depends on f, the strategies of all players. The benefit is associated with the cost $c_P(f)$, as this is the income a pathowner will receive from the flow units. (See Definition 2.4 for a detailed description of the benefit function.) The path player game is noncooperative and thus it is possible that the flow created by the decisions of the players is not feasible. For instance, if the benefit is an increasing function each player will try to get as much flow as possible and as a consequence the flow rate could be exceeded. A penalty is introduced to avoid an infeasible flow. In the case of infeasibility, the benefit of each player will be $-M$, with $M > 0$ being sufficiently large. We show in Section 2.2.3 that an infeasible flow may also be an equilibrium situation.

Our model also incorporates a social aspect: the community of players cares for players that receive only a small flow. These players shall get at least a fixed minimum income. If the flow of a player P lies below the security limit $\omega_P \geq 0$, she will receive a fixed security payment $\kappa_P > -M$.

Definition 2.3. *A path* P *is called* underloaded, *if* $f_P < \omega_P$ *and* loaded *otherwise.*

---

[1] Note that our definition of feasible flow differs from the definition in the routing game literature, where feasibility is obtained if the flow meets the flow rate exactly.
[2] In the course of this chapter we denote both the path and the corresponding player by P, as both these notations are handled equivalently.

For positive κ_P, the security limit and payment serve as insurance that guarantees a fixed income for each player. In Section 2.2.4 the so-called edge-sharing effect is described: it is possible that competitors force a player into a situation where routing more flow means less benefit. In this case, the security limit can be a protection against this harmful behavior of the competitors and thus it satisfies the idea of a social protection.

On the other hand, if $\kappa_P < 0$ holds, the payment becomes a punishment for underloaded paths. Hence, the security payment represents the additional costs for maintaining an unused resource.

The benefit function $b_P(f)$ is first defined in a general way. For this general benefit, we prove the existence of an equilibrium in Section 2.2.3. To obtain more results, we restrict the benefit function later on to special cases. Let $\mathbf{0}_n = (0,\ldots,0)^T$ be the vector that contains n times the entry 0. Summarizing, we have the following definition of the benefit function $b_P(f)$.

Definition 2.4. *The* benefit function *of player $P \in \mathcal{P}$ in a path player game is given for $f \geq \mathbf{0}_{|\mathcal{P}|}$ and $\kappa_P > -M$ as*

$$
b_P(f) = \begin{cases} c_P(f) & \text{if } \sum_{P_k \in \mathcal{P}} f_{P_k} \leq r \wedge f_P \geq \omega_P \\ \kappa_P & \text{if } \sum_{P_k \in \mathcal{P}} f_{P_k} \leq r \wedge f_P < \omega_P \\ -M & \text{if } \sum_{P_k \in \mathcal{P}} f_{P_k} > r \end{cases}
$$

where $c_P(f) = \sum_{e \in P} c_e(f_e)$, as given in Definition 2.1.

Apart from nonnegativity of ω_P, we have made no assumptions on ω_P and κ_P. The values are independent of each other for the time being. Nevertheless, it could be meaningful to define these values such that the security payment for the underloaded paths is covered by the security penalty (if existing) of the loaded paths. In the case of infeasibility, the players are punished with a penalty of $-M$. In all other cases, the benefit equals $c_P(f)$, the sum of cost functions $c_e(f_e)$ over all edges that belong to the path. The only assumption we have imposed on the cost functions c_e is that they have to be continuous and nonnegative for nonnegative flows f_e on all edges $e \in E$.

2.1.4 Game Types

In the following, we present a survey of the restrictions on game types we investigate in this work. The specifications are taken with regard to the network topology, cost functions, and general ones.

General Specifications
No security limit

Here, each player has a zero security limit; that is, $\omega_P = 0 \,\forall\, P \in \mathcal{P}$. If the flow is feasible, each player will receive $c_P(f_P)$ as the benefit. In this case, the payments κ_P are meaningless, as they will never be

used, hence no security limit causes no security payment. In games
with $\omega_P = 0$, we therefore set $\kappa_P = 0$.

No security limit is assumed for various results in Section 2.2.5. It is
also necessary for some results in Section 2.3 and is generally assumed
in Section 2.4.

Noncompensative security property

If a game has a noncompensative security (NCS) property (see Section 2.2.4 for the definition), each player will have incentive to route a
flow greater than or equal to the security limit ω_P, if this is possible.
Thus, a player will only take advantage of the security payment if she
is forced to. Games with this property will turn out to have nice behavior in terms of equilibria for strictly increasing cost functions; see
Section 2.2.5.

Trivial game

A game is called trivial if the sum of the security limits ω_P exceeds
the flow rate, that is, if $\sum_{P \in \mathcal{P}} \omega_P > r$. In these games, it may happen
that the flow rate r is completely used, and every player routes a flow
less than his security limit. See Section 2.2.4 for details. Nontriviality
is required to prove a necessary and sufficient condition for equilibria
in games with strictly increasing costs in Section 2.2.5.

Two-player game

For games on polyhedra (see Chapter 3), which are played by two players, the set of equilibria in the original game and the set of equilibria
in the extension of the game to the hypercuboid coincide.

Path player game (PPG) property

This property is interesting for games on polyhedra (see Chapter 3),
which are a generalization of path player games. If a game on a polyhedron satisfies a PPG property, the existence of potential functions
and the existence of equilibria are proved in Section 3.4.

Specifications of Network Topology

Path-disjoint network

In path-disjoint networks, no path shares an edge with another path
(see Section 2.2.4 for the definition). In this case, some negative effects
in the behavior of the benefit functions can be prevented, for instance,
the edge-sharing effect described in Section 2.2.4. As a consequence,
path-disjoint networks are useful to obtain games with the NCS property (see Section 2.2.4).

Regarding dominance, we show in Section 2.3.2 that in path-disjoint
networks, nondominated flows are equilibria. In addition to strictly increasing cost functions, the equality of the set of nondominated flows
and the set of equilibria holds.

Specifications of Cost Functions

Strictly increasing cost functions

In path player games, we are able to present for strictly increasing costs a necessary condition for a flow f^* to be an equilibrium. If the game has, in addition, a no security limit, or if the NCS property together with nontriviality is satisfied, we obtain a necessary and sufficient condition; see Section 2.2.5.

In terms of dominating flows (see Definition 2.68), we show in Section 2.3.2 that for strictly increasing costs, the set of nondominated flows is contained in the set of equilibria. If the game network is in addition path-disjoint, even equality of these two sets holds.

Regarding computation of equilibria, we show in Section 2.4.5 that a greedy approach delivers an equilibrium in a finite number of steps.

In games on polyhedra (Chapter 3), we show for strictly increasing cost functions that equilibria exist (which is not given for games on polyhedra in general) and we present a full characterization of the set of equilibria.

Differentiable cost functions

In path player games with differentiable cost functions, benefit functions are still nondifferentiable. A quasi-derivative is introduced, and a necessary condition for equilibria is developed in Section 2.2.5.

Differentiable and concave cost functions

Path player games with differentiable and concave cost functions are a special instance of the case described before. If we assume no security limit in addition, we obtain a necessary and sufficient condition for equilibria in Section 2.2.5.

Convex cost functions

In path player games with convex cost functions and no security limit, we provide in Section 2.2.5 a sufficient and necessary condition for equilibria, by which we can describe a dominating strategy set.

Regarding computation of equilibria, we show in Section 2.4.5 that a greedy approach delivers an equilibrium in a finite number of steps.

For games on polyhedra with convex costs, it is proved in Section 3.2 that equilibria, if they exist, have to lie on the boundary of the polyhedron.

Linear cost functions

In path player games, we show for linear cost functions that computation of equilibria is possible by using a greedy approach, which delivers an equilibrium in a finite number of steps (see Section 2.4.5).

In games on polyhedra with linear costs, equilibria exist. Furthermore, the set of equilibria can be described by the solution set of a linear program; see Section 3.2. In Section 3.4 it is proved that games on polyhedra with linear costs are potential games.

2.1.5 Topology of Networks in Path Player Games

In this section, we describe the concept of a *standard network* $G(n)$ which is able to represent all possible networks for a path player game with n players. The idea is to reduce any network of a path player game to a standard network, without changing the structure of the costs and thus without changing equilibria. With this technique of reducing any network, we are able to compute and study equilibria in the standard network which is planar and consists of $2^n - 1$ edges. Although these results are useful for path player games on arbitrary networks and are also interesting for the introduction of the path player game(PPG) property in Section 3.4.1, standard networks have no impact on the remaining chapters in this text. A hurried reader may skip this section without losing the scope.

Denote with $\mathbb{P}(n)$ the power set of the set of players $\{P_1, \ldots, P_n\}$. Furthermore, $\mathbb{P}(n, k) = \{m \in \mathbb{P}(n) : |m| = k\}$ is the set of elements in $\mathbb{P}(n)$ that have k entries. Note that

$$|\mathbb{P}(n, k)| = \binom{n}{k}$$

holds. Consider edges that are owned by k players. There are $\binom{n}{k}$ possibilities that k out of n players share an edge.

Definition 2.5. *For a set of players* $m \in \mathbb{P}(n)$, *the* class of edges *that are owned exactly by the set of players* m, *is given by*

$$e_m = \{e : \{P : e \in P\} = m\}, \quad m \in \mathbb{P}(n).$$

If we neglect the empty set $m = \emptyset$ (as each edge is owned at least by one path), we have $2^n - 1$ different classes of edges. Note that the sets $e_m, m \in \mathbb{P}(n)$ make up a partition of the set of edges E. Note furthermore that e_m is not just the intersection of the paths $P \in m$. This can be verified in Example 2.7.

Example 2.6. In a path player game with $n = 4$ players, we have the following classes of edges.

$k = 1$	$e_{\{P_1\}}, e_{\{P_2\}}, e_{\{P_3\}}, e_{\{P_4\}},$
$k = 2$	$e_{\{P_1,P_2\}}, e_{\{P_1,P_3\}}, e_{\{P_1,P_4\}}, e_{\{P_2,P_3\}}, e_{\{P_2,P_4\}}, e_{\{P_3,P_4\}},$
$k = 3$	$e_{\{P_1,P_2,P_3\}}, e_{\{P_1,P_2,P_4\}}, e_{\{P_1,P_3,P_4\}}, e_{\{P_2,P_3,P_4\}},$
$k = 4$	$e_{\{P_1,P_2,P_3,P_4\}}.$

Classes of edges e_m may be empty, see the following example.

Example 2.7. Consider the game illustrated by Figure 2.1 with

$P_1 = \{1, 4, 6\},$
$P_2 = \{2, 3, 4, 6\},$
$P_3 = \{5, 6\}.$

The following classes of edges are given in this game.

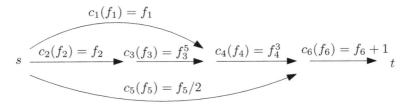

Fig. 2.1. Game network of Example 2.7.

$k = 1$ $e_{\{P_1\}} = \{1\}$, $e_{\{P_2\}} = \{2,3\}$, $e_{\{P_3\}} = \{5\}$,

$k = 2$ $e_{\{P_1,P_2\}} = \{4\}$, $e_{\{P_1,P_3\}} = \emptyset$, $e_{\{P_2,P_3\}} = \emptyset$,

$k = 3$ $e_{\{P_1,P_2,P_3\}} = \{6\}$.

For n, $k \in \mathbb{N}, k \leq n$, the following is true;

$$|\mathbb{P}(n,k)| = |\mathbb{P}(n, n-k)|.$$

Definition 2.8. *For each $m \in \mathbb{P}(n)$ we define the* complement *of m by*

$$\bar{m} = \{P_1, \ldots, P_n\} \setminus m.$$

It can be observed that for for each m, the complement $\bar{m}$ is well defined, and that for $m \in \mathbb{P}(n,k)$ it holds that $\bar{m} \in \mathbb{P}(n, n-k)$. Furthermore, for all pairs $m, \bar{m}$ it holds that

$$m \cap \bar{m} = \emptyset \ \wedge \ m \cup \bar{m} = \{P_1, \ldots, P_n\}. \tag{2.1}$$

In the following, we describe the transformation of the network, of the paths, and of the costs assigned to edges. Afterwards, we introduce the transformation of the flow and show that the benefit is not changed by this transformation. We start with Algorithm 1, which creates the standard network. This algorithm generates a network consisting of $2^n - 1$ edges $\hat{e}_m$, one for each class of edges e_m. These edges are connected by $2^{n-1} + 1$ vertices (including source and sink). The source is followed by the single edge $\hat{e}_{\{P_1,\ldots,P_n\}}$ referring to the class $e_{\{P_1,\ldots,P_n\}}$. After adding a vertex to that edge, a pair e_m, $e_{\bar{m}}$ is attached as parallel edges. This is repeated until all pairs e_m, $e_{\bar{m}}$ are attached sequentially, connected by vertices. Finally, the sink vertex terminates this sequence.

Definition 2.9. *For a path player game played with n players, we denote the standard network by $G(n) = (\hat{V}, \hat{E})$, where $G(n)$ is created by Algorithm 1. Furthermore, for all players $P_i \in \{P_1, \ldots, P_n\}$, we define $\hat{P}_i$, the path connecting s and t in the standard network, as the following. $\hat{P}_i$ is the sequence of edges $\hat{e}_m$ with $P_i \in m$, where the edges are given in the same order as they appear in the standard network $G(n)$.*

Algorithm 1 Generation of standard network $G(n)$

1: Create source vertex s
2: Attach edge $\hat{e}_{\{P_1,\dots,P_n\}}$ to s and attach a transit vertex, to terminate the edge.
3: $k := 1$
4: **while** $k \leq n/2$ **do**
5:     $\mathbb{P}^k := \mathbb{P}(n,k) \cup \mathbb{P}(n, n-k)$
6:     **while** $\mathbb{P}^k \neq \emptyset$ **do**
7:         Choose $m \in \mathbb{P}^k$
8:         Set $\bar{m} := \{P_1, \dots, P_n\} \setminus m$
9:         Attach edges $\hat{e}_m$ and $\hat{e}_{\bar{m}}$ as parallel edges to the transit vertex of the previous step and terminate with another transit vertex
10:       Set $\mathbb{P}^k := \mathbb{P}^k \setminus \{m, \bar{m}\}$
11:     **end while**
12:     $k := k + 1$
13: **end while**
14: The vertex attached in the previous iteration is labeled as sink vertex t.

Note that $\hat{P}_i$ is existing and well defined, as after passing any vertex (apart from the source) there is a parallel pair of edges $m, \bar{m}$, where each player P_i is assigned to exactly one of these edges. On the other hand, for every $m \in \mathbb{P}(n)$ there is an edge $\hat{e}_m$ in $G(n)$. Thus, each combination of players sharing edges is possible.

Example 2.10. Figure 2.2 illustrates the standard network for $n = 4$.

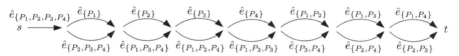

Fig. 2.2. Standard network $G(4)$.

Definition 2.11. *For a standard network $G(n) = (\hat{V}, \hat{E})$, the cost on an edge is given by*

$$c_{\hat{e}_m}(f_{\hat{e}_m}) = \sum_{e \in e_m} c_e(f_{\hat{e}_m}).$$

If $e_m = \emptyset$ holds for an m, the cost on the corresponding edge $\hat{e}_m$ reduces to $c_{\hat{e}_m}(f_{\hat{e}_m}) = 0$.

Definition 2.12. *We call the game based on the standard network $G(n)$ the path player game in* standard form.

Next, we describe how the flow changes because of the transformation to standard form. In particular, the flow on the paths stays the same, but as the edges have been transformed, the flow on the edges is adjusted. Moreover, we prove that the benefit, and thus the equilibria, are not changed.

Definition 2.13. *The flow in a path player game in standard form is given by*

$$f_{\hat{P}} = f_P.$$

Lemma 2.14. *For a path player game in standard form, the flow on an edge is given by*

$$f_{\hat{e}_m} = \sum_{P \in m} f_P.$$

Proof.

$$f_{\hat{e}_m} = \sum_{\hat{P}: \hat{e}_m \in \hat{P}} f_{\hat{P}} = \sum_{P \in m} f_{\hat{P}} = \sum_{P \in m} f_P. \qquad \square$$

Note that $f_e = \sum_{P: e \in P} f_P = f_{\hat{e}_m}$ holds for all $e \in e_m$.

Lemma 2.15. *Consider a path player game Γ on a network $G = (V, E)$ with n players. The corresponding path player game in standard form is equivalent to Γ.*

Proof. In the standard network it holds for each player P:

$$\begin{aligned}
c_{\hat{P}}(f) &= \sum_{\hat{e}_m \in \hat{P}} c_{\hat{e}_m}(f_{\hat{e}_m}) \\
&= \sum_{m: P \in m} c_{\hat{e}_m}(f_{\hat{e}_m}) = \sum_{m: P \in m} \sum_{e \in e_m} c_e(f_e) \qquad (2.2) \\
&= \sum_{e \in P} c_e(f_e) = c_P(f).
\end{aligned}$$

All other components of the benefit function, such as security limit and payment, flow rate and infeasibility penalty, are not dependent on the edges e. Hence, the benefit $b_P(f)$ is not changed by the modification. $\square$

Example 2.16. For the game presented in Example 2.7 on page 15, the standard network $G(3)$ and the corresponding cost functions are illustrated in Figure 2.3.

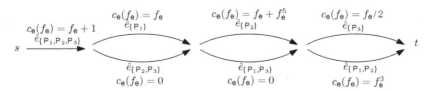

Fig. 2.3. Standard network $G(3)$ for Example 2.16.

2.2 Equilibria in Path Player Games

2.2.1 Introduction

In this section, we analyze *equilibria* in path player games with a general benefit function and for special instances of benefit functions. Our definition of an equilibrium follows the definition of an equilibrium in the sense of Nash (see e.g., [Owe95]). An equilibrium is a game situation where none of the players is able to obtain a better outcome by unilaterally changing her strategy. In other words, assuming that competitors keep their current strategies, a player will not be able to improve her benefit. Hence, in an equilibrium none of the players has a reason to change the chosen strategy. This situation characterizes a stable state of the system.

Subsequent to this introduction, we define in Section 2.2.2 the best reaction set for the players P and other notations required for the analysis of equilibria. We consider equilibria with respect to feasible and infeasible flows in Section 2.2.3. The existence of feasible equilibria is proved by a fixed-point argument. The presented existence proof induces even the existence of feasible equilibria in pure strategies (see Remark 2.27). Infeasible equilibria exist and their characteristics are given below.

We present special instances of path player games and describe their properties in Section 2.2.4. For these types of games, including special benefit functions, we obtain more results about equilibria in Section 2.2.5.

2.2.2 The One-Dimensional Benefit Function

For determining equilibria in the path player game, we take into account the reasoning of the players. We consider the benefit of a single player who only has the possibility to change his own strategy, whereas the strategies of the competitors are fixed. The single player has to solve a one-dimensional optimization problem, depending on the flow f_P. For this purpose, it is helpful to define a one-dimensional benefit function for each player P.

We first consider the cost function $c_P(f)$, as it is the most important element of the benefit. We define $f_{-P} \in \mathbb{R}^{|\mathcal{P}|-1}$ by deleting the component f_P from f belonging to path P. If we just want to consider the influence of f_P, we fix the strategies f_{-P} of the competitors and analyze the cost depending on f_P.

Definition 2.17. *The* one-dimensional cost function *assigned to a player* $P \in \mathcal{P}$ *for a given flow* f_{-P} *and for* $f_P \geq 0$ *is denoted by*

$$\tilde{c}_P(f_P) = c_P(f_{-P}, f_P) = \sum_{e \in P} c_e \left(f_P + \sum_{P_k \in \mathcal{P} \setminus \{P\}: e \in P_k} f_{P_k} \right).$$

The one-dimensional cost function $\tilde{c}_P(f_P)$ only depends on the one-dimensional f_P once f_{-P} is fixed. The term

$$\sum_{P_k \in \mathcal{P} \setminus \{P\} : e \in P_k} f_{P_k}$$

is constant for fixed f_{-P}.

Lemma 2.18. *If $c_e(f_e)$ is a convex (concave) function, then $\tilde{c}_P(f_P)$ is also a convex (concave) function.*

Proof. The proof is done for convex functions. The statement for concave functions can be shown analogously by using the fact that a function $-f(x)$ is convex if $f(x)$ is concave.

Let $c_e(x)$ be a convex function and consider the constant $C_{e,P} \in \mathbb{R}$. It holds that $c_e(x + C_{e,P})$ is a convex function. It can be shown that the sum of convex functions is again a convex function, such that $\sum_{e \in P} c_e(x + C_{e,P})$ is also convex in x. It follows immediately that $\tilde{c}_P(f_P)$ is a convex function.   □

If the strategies of all players (except P) are fixed, we introduce the following.

Definition 2.19. *The* one-dimensional benefit *for a player P and a flow $f_P \geq 0$ with respect to a given flow f_{-P} is denoted by*

$$\tilde{b}_P(f_P) = b_P(f_{-P}, f_P).$$

The main difference from the benefit function introduced in Definition 2.4 is that $\tilde{b}_P(f_P)$ depends on a scalar f_P and not on a vector f. The one-dimensional benefit shows the player what she is able to achieve in the current situation if all other players keep their chosen strategies.

Definition 2.20. *The* decision limit *of player P with respect to a given flow f_{-P} is denoted by*

$$d_P(f_{-P}) = r - \sum_{P_k \in \mathcal{P} \setminus \{P\}} f_{P_k}.$$

If no confusion regarding the chosen f_{-P} arises, we denote the decision limit just by d_P. The interval $[0, d_P]$ is called the decision interval *of player P.*

The decision interval indicates the set of feasible strategies for P. Player P is allowed to choose any nonnegative value for f_P, but he should choose from $[0, d_P]$ as for larger values the system will fall into infeasibility and he will get a penalty of $-M$.

From the definition of the decision limit we obtain the next corollary.

Corollary 2.21. *Any flow f satisfies*

$$\exists P_k : f_{P_k} = d_{P_k} \quad \Rightarrow \quad \forall P \in \mathcal{P} : f_P = d_P.$$

Using the one-dimensional cost function $\tilde{c}_P(f_P)$ defined above and the decision limit d_P, the following corollary describes the one-dimensional benefit function in more detail.

Corollary 2.22. *In a path player game and for a given flow f_{-P}, the one-dimensional benefit for a player P and for $f_P > 0$ is*

$$\tilde{b}_P(f_P) = \begin{cases} \tilde{c}_P(f_P) & if\ f_P \leq d_P \wedge f_P \geq \omega_P \\ \kappa_P & if\ f_P \leq d_P \wedge f_P < \omega_P \\ -M & if\ f_P > d_P \end{cases} .$$

Figure 2.4 illustrates an example for a one-dimensional benefit function. The function $\tilde{b}_P(f_P)$ is characterized by three parts: the two constant regions generated by the security payment κ_P, the infeasibility penalty $-M$, and the middle part, created by the cost function $\tilde{c}_P$. As each player will try to maximize his benefit, the next definition is useful.

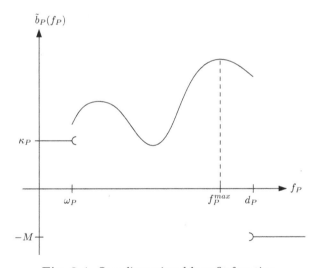

Fig. 2.4. One-dimensional benefit function.

Definition 2.23. *Given a flow f_{-P}, the* best reaction set *for a player P is*

$$f_P^{\max}(f_{-P}) = \left\{ f_P \geq 0 : f_P\ maximizes\ \tilde{b}_P(f_P) \right\}.$$

If no confusion regarding the chosen f_{-P} arises, we denote the best reaction set by $f_P^{\max}$.

The set $f_P^{\max}$ indicates the flows that will maximize player P's benefit if the other players choose to play f_{-P}. Thus, if player P chooses a flow from $f_P^{\max}$, she chooses a best reaction to the actions of her competitors. The next lemma gives a sufficient condition to ensure nonemptiness of this set.

Lemma 2.24. *Consider a path player game with cost functions c_e being continuous for all edges $e \in E$. Then, the sets $f_P^{\max}$ are nonempty for all $P \in \mathcal{P}$.*

Proof. Consider the intervals $I_1 = [0, \omega_P)$, $I_2 = [\omega_P, d_P]$, and $I_3 = (d_P, \infty)$. Because $\tilde{b}_P(f_P)$ is constant on I_1 and I_3, maxima exist for these intervals. The existence of a maximum on I_2 is confirmed by the Weierstrass extreme value theorem because $\tilde{b}_P(f_P)$ is continuous on I_2 and I_2 is compact. The maximum of these three single maxima hence is the overall maximum. $\square$

As we assume in general having continuous cost functions, the lemma holds for all game instances observed in this text.

Remark 2.25. In our game, we assume having continuous costs. Otherwise, for noncontinuous cost functions c_e, we could not guarantee the existence of a benefit maximizing flow. We can easily construct instances where the jump of the function allows only determining a supremum, but not a maximum; see, for instance, Figure 2.5.

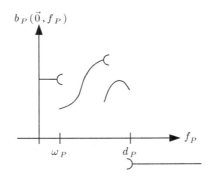

Fig. 2.5. Best reaction set $f_P^{\max}$ is empty.

2.2.3 Equilibria for General Benefits and Existence of Equilibria

After modeling the benefit of the single player in a game situation by the one-dimensional benefit, we are now able to define equilibria in a path player game.

Definition 2.26. *In a path player game a flow f^* is an equilibrium if and only if for all players $P \in \mathcal{P}$ and for all $f_P \geq 0$ the following holds,*

$$b_P(f_{-P}^*, f_P^*) \geq b_P(f_{-P}^*, f_P).$$

Remark 2.27. The definition of an equilibrium in path player games is given for the choice of *pure strategies*. To choose a pure strategy means that a player P decides for a determined strategy $f_P \in \mathbb{R}_+$. In contrast, considering *mixed strategies* means that a player chooses a probability distribution over his strategy set. As a result of playing mixed strategies, *expected payoffs*, or expected benefits, are obtained. Mixed strategies are of relevance, for example, in finite n-person games, where mixed-strategy equilibria do exist, but equilibria in pure strategies are not given in general; see [Owe95]. We do not consider mixed strategies in this analysis. Later on, we prove the existence of pure-strategy equilibria, which supports the restriction to pure strategies.

According to this definition, an equilibrium is a state where no player can obtain a better outcome by changing her strategy, if the competitors stay with their chosen strategy. An equilibrium is a stable state in which none of the selfish users has an incentive to change behavior. Although this state is stable it is not necessarily a good solution for the players. It may be possible to find other solutions, equilibria and even nonequilibria, that are better for some players, without being worse for any player. We call such equilibria *dominated* and investigate this aspect in more detail in Section 2.3.

Best reaction sets are useful for a different characterization of equilibria. A flow is an equilibrium if each player chooses a best reaction strategy to the strategies of his opponents.

Corollary 2.28. *In a path player game a flow f^* is an equilibrium if and only if for all $P \in \mathcal{P}$ with respect to f^*_{-P} it holds that*

$$f_P^* \in f_P^{\max}(f_{-P}^*).$$

Proof. Assume that f_P^* is in $f_P^{\max}(f_{-P}^*)$ for all $P \in \mathcal{P}$.

$$\Leftrightarrow f_P^* \in \{f_P \geq 0 : f_P \text{ maximizes } \tilde{b}_P(f_P)\} \quad \forall\, P \in \mathcal{P}$$

$$\Leftrightarrow \tilde{b}_P(f_P^*) \geq \tilde{b}_P(f_P) \qquad \forall\, P \in \mathcal{P} \wedge \forall\, f_P \geq 0$$

$$\Leftrightarrow b_P(f_{-P}^*, f_P^*) \geq b_P(f_{-P}^*, f_P) \qquad \forall\, P \in \mathcal{P} \wedge \forall\, f_P \geq 0$$

$$\Leftrightarrow f^* \text{ is an equilibrium.} \qquad\qquad \square$$

In path player games two types of equilibria are distinguished: equilibria flows that are feasible and those that are infeasible. These two variants are discussed in the following paragraphs.

Feasible Equilibria

In this section, we define feasible equilibria and prove their existence for games with continuous cost functions.

Definition 2.29. *In a path player game, a flow* f^* *is called a* feasible equilibrium *if* f^* *is a feasible flow and an equilibrium.*

The next corollary follows immediately from the definition of an equilibrium (Definition 2.26).

Corollary 2.30. *A flow* f^* *is a feasible equilibrium in a path player game if and only if* f^* *is feasible and for all paths* P *in* $\mathcal{P}$ *and for all* $f_P \leq d_P$ *it holds that*

$$\tilde{b}_P(f_P^*) \geq \tilde{b}_P(f_P).$$

For infinite games with continuous benefits it is known that there exist equilibria in mixed strategies (see Remark 2.27) if the strategy spaces are nonempty and compact. Even more, if we assume continuous and quasi-concave benefit functions, there exist pure strategy equilibria (see [FT91]). In our game, we cannot assume continuous benefit functions, but we have benefit functions where we can guarantee a nonempty best reaction set $f_P^{\max}$. This allows us to prove the existence of an equilibrium according to Definition 2.26, that is, an equilibrium in pure strategies.[3] We can even show that there is always a feasible equilibrium. Let

$$\mathbb{F} = \left\{ f \in \mathbb{R}_+^{|\mathcal{P}|} : \sum_{P \in \mathcal{P}} f_P \leq r \right\} \tag{2.3}$$

be the *set of feasible flows* f. Note that $\mathbb{F}$ is closed, bounded, and convex.

Theorem 2.31 (Existence of feasible equilibria). *In a path player game with continuous cost functions* c_e *for all edges* $e \in E$, *a feasible equilibrium exists.*

Proof. Consider the closed, bounded, and convex set of feasible flows $\mathbb{F}$. Furthermore, consider the mapping $f' = T(f)$ with the components $f_P' = t(f_P)$ given by

$$f_P' = f_P + \begin{cases} \min \left\{ f_P^m - f_P; \dfrac{f_P^m - f_P}{\sum_{P_k \in \mathcal{P}: f_{P_k} < f_{P_k}^m} (f_{P_k}^m - f_{P_k})} \cdot d \right\} & \text{if } f_P < f_P^m \\ f_P^m - f_P & \text{if } f_P \geq f_P^m \end{cases} , \tag{2.4}$$

where $f_P^m = \min \{f_P^{\max}\}$ is chosen as the smallest flow that is benefit maximizing[4] and $d = r - \sum_{P \in \mathcal{P}} f_P$ is the remaining flow that can be distributed among the players maintaining feasibility. Note that it holds for all $P \in \mathcal{P}$ that

---

[3] We do not stress the fact that an equilibrium is in pure strategies any more, as we are considering only pure strategy equilibria in this analysis, which is also reflected by our definition of an equilibrium (Definition 2.26).

[4] Note that the proof would also hold for a different definition of choosing $f_P^m \in f_P^{\max}$, as long as f_P^m is well defined.

$$d = d_P - f_P \geq 0.$$

Note furthermore that, by Lemma 2.24, f_P^m exists and, by definition of $f_P^{\max}$, it holds that

$$0 \leq f_P^m \leq d_P = r - \sum_{P_k \in \mathcal{P} \setminus \{P\}} f_{P_k}. \tag{2.5}$$

(See page 27 for a visualization and interpretation of the mapping T.) In the following we prove that T is a continuous mapping of $\mathbb{F}$ into itself. Then, by Brouwer's fixed point theorem, a fixed point $f = T(f)$ exists in $\mathbb{F}$. Finally, we show that each fixed point in $\mathbb{F}$ represents an equilibrium. In this way, we are able to guarantee the existence of a feasible equilibrium.

Part (a) $T : \mathbb{F} \to \mathbb{F}$.

Denote the sets $\mathcal{P}_1 = \{P \in \mathcal{P} : f_P < f_P^m\}$ and $\mathcal{P}_2 = \{P \in \mathcal{P} : f_P \geq f_P^m\}$. Then:

$$f_P' = \overbrace{f_P}^{\geq 0} + \begin{cases} \min\left\{ \overbrace{f_P^m - f_P}^{\geq 0}; \overbrace{\dfrac{f_P^m - f_P}{\sum_{P_k \in \mathcal{P}_1}(f_{P_k}^m - f_{P_k})} \cdot d}^{\geq 0} \right\} & \text{if } f_P < f_P^m \\[4mm] \underbrace{f_P^m - f_P}_{\geq -f_P} & \text{if } f_P \geq f_P^m \end{cases}$$

$$\Rightarrow f_P' \geq 0 \; \forall \; P \in \mathcal{P}. \tag{2.6}$$

$$\sum_{P \in \mathcal{P}} f_P' = \sum_{P \in \mathcal{P}_1} \left(f_P + \min\left\{ f_P^m - f_P; \frac{f_P^m - f_P}{\sum_{P_k \in \mathcal{P}_1}(f_{P_k}^m - f_{P_k})} \cdot d \right\} \right)$$

$$+ \sum_{P \in \mathcal{P}_2} (f_P + f_P^m - f_P)$$

$$= \sum_{P \in \mathcal{P}} f_P + \sum_{P \in \mathcal{P}_1} \min\left\{ f_P^m - f_P; \frac{f_P^m - f_P}{\sum_{P_k \in \mathcal{P}_1}(f_{P_k}^m - f_{P_k})} \cdot d \right\}$$

$$+ \overbrace{\sum_{P \in \mathcal{P}_2} (f_P^m - f_P)}^{\leq 0}$$

$$\leq \sum_{P \in \mathcal{P}} f_P + \sum_{P \in \mathcal{P}_1} \frac{f_P^m - f_P}{\sum_{P_k \in \mathcal{P}_1}(f_{P_k}^m - f_{P_k})} \cdot d$$

$$= \sum_{P \in \mathcal{P}} f_P + d = \sum_{P \in \mathcal{P}} f_P + r - \sum_{P \in \mathcal{P}} f_P$$

$$= r.$$

As from (2.6) and $\sum_{P \in \mathcal{P}} f'_P \leq r$ we obtain $f' \in \mathbb{F}$.

Part (b) $T(f)$ is continuous.

Consider the following cases.

(i) $f_P > f_P^m$

$f'_P = f_P^m \vee f_P > f_P^m$; that is, $t(f_P)$ is continuous.

(ii) $f_P = f_P^m + 0$

$f'_P = f_P^m$ for $f_P = f_P^m + 0$; tht is, $t(f_P)$ is upper semicontinuous at $f_P = f_P^m + 0$.

(iii) $f_P < f_P^m$

Consider $g(f) = f_P^m - f_P$ and

$$h(f) = \frac{f_P^m - f_P}{\sum_{P_k \in \mathcal{P}_1} (f_{P_k}^m - f_{P_k})} \cdot d.$$

The functions $g(f)$ and $h(f)$ are continuous and so the minimum of both functions is continuous too. It follows that $t(f_P)$ with $f'_P = f_P + \min\{g(f); h(f)\}$ is a continuous mapping.

(iv) $f_P = f_P^m - 0$

Consider the marginal value of the mapping that we take for each flow f where $f_P \to f_P^m - 0$:

$$\lim_{f : f_P \to f_P^m - 0} \left(\overbrace{f_P}^{\to f_P^m} + \min \left\{ \overbrace{f_P^m - f_P}^{\to 0}; \overbrace{\frac{\overbrace{f_P^m - f_P}^{\geq 0}}{\sum_{P_k \in \mathcal{P}_1} (f_{P_k}^m - f_{P_k})} \cdot \overbrace{d}^{\geq 0}}^{\geq 0} \right\} \right) = f_P^m.$$

Thus, $t(f_P)$ is lower semicontinuous at $f_P = f_P^m - 0$.

Hence, T is continuous.

Part (c) $f = T(f) \Rightarrow f$ is an equilibrium.

Consider the mapping (2.4) and rewrite it as $f'_P = f_P + K_P$, where K_P substitutes the parenthesis. From $f = T(f)$ follows $f'_P = f_P$ and thus $K_P = 0$ for all $P \in \mathcal{P}$. Consider two cases:

(i) $f_P < f_P^m$

As $K_P = 0$ and $f_P^m - f_P > 0$ it follows that $d = 0$ has to hold. From (2.5) we get:

$$0 = d = r - \sum_{P \in \mathcal{P}} f_P \geq f_P^m - f_P \quad \Rightarrow \quad f_P \geq f_P^m,$$

which contradicts the assumption; that is, $\nexists P : f_P < f_P^m$.

(ii) $f_P \geq f_P^m$

From $K_P = 0$ we obtain $f_P^m - f_P = 0$. It follows that $f_P = f_P^m \in f_P^{max}$.

We conclude that $f_P \in f_P^{max} \ \forall \ P \in \mathcal{P}$; that is, f is an equilibrium. $\qquad \square$

Note that Nash's proof of the existence of a mixed strategy equilibrium in finite games (see [Nas50]) also uses a fixed-point argument, but with a different mapping $T(f)$. Two more versions of existence proofs are presented in Theorem 2.103 (page 73) and Theorem 2.118 (page 84), where further results about potential functions in path player games allow different approaches to prove the existence of equilibria.

Figures 2.6 and 2.7 illustrate the mapping T. The mapping can be interpreted as a simple auction where the players bid the flow they want to route over their path. In particular, each player asks to receive the flow f_P^m. Then, each player receives a flow f_P' which depends on all bids and on the amount of flow that can be distributed without exceeding the flow rate r. If the current flow of a player P is greater than or equal to f_P^m, then she is given exactly $f_P' = f_P^m$, as reducing flow will not violate the flow rate. If $f_P < f_P^m$ holds (i.e., P will ask for a larger flow), we have to distinguish two cases. The first case is illustrated in Figure 2.6. Here,

$$\sum_{P_k \in \mathcal{P}_1} (f_{P_k}^m - f_{P_k}) > d$$

holds; that is, the players want to increase their flow, but ask for more flow than available. In this case the flow rate would be violated if each player received his bid. Hence, each player receives a fraction of d proportional to his bid and smaller than his bid. In the second case, the sum of the players' bids does not exceed r: $\sum_{P_k \in \mathcal{P}_1} (f_{P_k}^m - f_{P_k}) \leq d$. Each player will receive exactly his bid, which is illustrated in Figure 2.7.

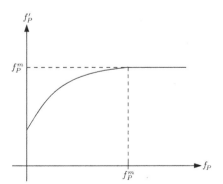

Fig. 2.6. Players receive flow smaller than bids.

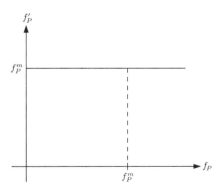

Fig. 2.7. Players receive exact bids as flow.

Infeasible Equilibria

In the course of this text we concentrate on feasible equilibria. Nevertheless, in a path player game it is possible that an infeasible flow is an equilibrium. Analogously to Definition 2.29 we call such a flow an *infeasible equilibrium*. An infeasible flow f is an infeasible equilibrium if for each player $P \in \mathcal{P}$ the following holds: f will stay infeasible, even when any $P \in \mathcal{P}$ decides to send a zero-flow. That means none of the players has the chance to create a feasible flow by acting unilaterally. As nobody can improve the benefit on his own, the observed flow is an equilibrium. In this case all players $P \in \mathcal{P}$ will be confronted with a constant benefit function $\tilde{b}_P(f_P) = -M$ for all $f_P \geq 0$. So the best reaction set will be $f_P^{\max} = [0, \infty)$ for all P and the condition of an equilibrium described in Lemma 2.28 is satisfied. In the case of infeasibility,

$$\sum_{P_k \in \mathcal{P} \setminus \{P\}} f_{P_k} > r$$

holds for all paths $P \in \mathcal{P}$. Thus, for the decision limit

$$d_P = r - \sum_{P_k \in \mathcal{P} \setminus \{P\}} f_{P_k} < 0$$

will hold for all players $P \in \mathcal{P}$. The decision intervals are then empty for all players.

The following lemma provides a complete characterization of infeasible equilibria.

Lemma 2.32. *In a path player game a flow f is an infeasible equilibrium if and only if for all paths P in $\mathcal{P}$ the following is satisfied.*

$$\sum_{P \in \mathcal{P}} f_P \geq r + \max_{P \in \mathcal{P}} f_P.$$

Proof.
Part (a) $\sum_{P \in \mathcal{P}} f_P \geq r + \max_{P \in \mathcal{P}} f_P \Rightarrow f$ is an infeasible equilibrium.
Consider a flow f such that $\sum_{P \in \mathcal{P}} f_P \geq r + \max_{P \in \mathcal{P}} f_P$ holds. This flow is infeasible due to $\max_{P \in \mathcal{P}} f_P > 0$. In addition for all paths P in $\mathcal{P}$ the following is true.

$$d_P = r - \sum_{P_k \in \mathcal{P} \setminus \{P\}} f_{P_k}$$

$$= r - \left(\sum_{P_k \in \mathcal{P}} f_{P_k} - f_P \right)$$

$$\leq r - \left(\overbrace{\sum_{P_k \in \mathcal{P}} f_{P_k}}^{\geq r} - \max_{P \in \mathcal{P}} f_P \right)$$

$$\leq 0$$

$$\Rightarrow \tilde{b}_P(f_P) = -M \ \forall \ f_P$$

$$\Rightarrow f_P^{\max} = [0, \infty).$$

We conclude that $f_P \in f_P^{\max} \ \forall \ P \in \mathcal{P}$ and hence, f is an equilibrium.

Part (b) f is an infeasible equilibrium $\Rightarrow \sum_{P \in \mathcal{P}} f_P \geq r + \max_{P \in \mathcal{P}} f_P$. Consider a flow f such that

$$\sum_{P \in \mathcal{P}} f_P > r \quad \text{and} \quad f_P \in f_P^{\max} \ \forall \ P \in \mathcal{P};$$

that is, f is an infeasible equilibrium. Assume that the claim is not true; that is,

$$\sum_{P \in \mathcal{P}} f_P < r + \max_{P \in \mathcal{P}} f_P.$$

Let $\bar{P}$ be such that $\max_{P \in \mathcal{P}} f_P = f_{\bar{P}}$. Then,

$$d_{\bar{P}} = r - \sum_{P \in \mathcal{P} \setminus \{\bar{P}\}} f_P = r - \left(\sum_{P \in \mathcal{P}} f_P - f_{\bar{P}} \right) = r - \overbrace{\left(\sum_{P \in \mathcal{P}} f_P - \max_{P \in \mathcal{P}} f_P \right)}^{< r}$$

$$\Rightarrow d_{\bar{P}} > 0$$

$$\Rightarrow \exists \ f_{\bar{P}}' : \tilde{b}_{\bar{P}}(f_{\bar{P}}') > -M$$

$$\Rightarrow f_{\bar{P}} \notin f_{\bar{P}}^{\max},$$

which contradicts the assumption and thus the claim follows. $\square$

As a result of the lemma just proved, for path player games, infinitely many infeasible equilibria exist.

2.2.4 Special Instances of Path Player Games

Path-Disjoint Network

Definition 2.33. *A set of paths $\mathcal{P}$ is called* disjoint *if for all pairs $P_1, P_2 \in \mathcal{P}$ with $P_1 \neq P_2$ it holds that $P_1 \cap P_2 = \emptyset$. We call a network* path-disjoint *if the set $\mathcal{P}$ of all paths from s to t is disjoint.*

Corollary 2.34. *In a path-disjoint network it holds for all $P \in \mathcal{P}$ that*

$$c_P(f) = c_P(f_{-P}, f_P) = c_P(\,\cdot\,, f_P);$$

that is, $c_P(f)$ depends only on f_P and is independent of f_{-P}.

Proof. As each edge is contained in not more than one path, the following holds.

$$c_P(f) = \sum_{e \in P} c_e(f_e) = \sum_{e \in P} c_e \left(\sum_{P:e \in P} f_P \right) = \sum_{e \in P} c_e(f_P) = c_P(\,\cdot\,, f_P). \qquad \square$$

Cost functions c_P with $c_P(f) = c_P(\,\cdot\,, f_P)$ are also known as *separable functions* (e.g., see [GY99]).

Corollary 2.35. *A path-disjoint network with n paths $P_1, \ldots, P_n$ connecting the source s and the sink t can be reduced to a simpler network with n parallel edges $e_1, \ldots, e_n$ connecting s and t with $\bar{c}_{e_i}(f_{e_i}) = c_{P_i}(f)$ for all $i = 1, \ldots, n$.*

Proof. It holds that

$$c_{P_i}(f) = c_{P_i}(\,\cdot\,, f_{P_i}) = \sum_{e \in P} c_e(f_{P_i}).$$

Choose $\bar{c}_{e_i} = \sum_{e \in P} c_e(f_{P_i})$ and the corollary follows. $\qquad \square$

Trivial Games

Definition 2.36. *We call a path player game with flow rate r and security limits ω_P trivial, if*

$$\sum_{P \in \mathcal{P}} \omega_P > r$$

holds, and nontrivial *otherwise.*

In trivial games, it is possible that in a situation where the complete flow rate r is used (i.e., $\sum_{P \in \mathcal{P}} f_P = r$) all players have made use of the security payment by setting $f_P < \omega_P$. In a nontrivial game there will be at least one player who is able to route a flow greater than or equal to ω_P without destroying feasibility. This observation is stated in the following lemma.

Lemma 2.37. *In a nontrivial path player game for any given feasible flow f there exists at least one $P \in \mathcal{P}$ such that $d_P \geq \omega_P$.*

Proof. Consider a nontrivial path player game; that is, $\sum_{P \in \mathcal{P}} \omega_P \leq r$ and a given feasible flow f. It holds for all $P \in \mathcal{P}$ that

$$d_P \quad = r - \sum_{P_k \in \mathcal{P} \setminus \{P\}} f_{P_k} = r - \sum_{P_k \in \mathcal{P}} f_{P_k} + f_P$$

$$\Rightarrow \sum_{P \in \mathcal{P}} d_P = |\mathcal{P}| \cdot r - |\mathcal{P}| \cdot \sum_{P \in \mathcal{P}} f_P + \sum_{P \in \mathcal{P}} f_P$$

$$= |\mathcal{P}| \cdot r - (|\mathcal{P}| - 1) \cdot \sum_{P \in \mathcal{P}} f_P$$

$$= r + (|\mathcal{P}| - 1) \cdot \overbrace{\left(r - \sum_{P \in \mathcal{P}} f_P \right)}^{\geq 0}$$

$$\Rightarrow \quad \sum_{P \in \mathcal{P}} d_P \geq r \geq \sum_{P \in \mathcal{P}} \omega_P$$

$$\Rightarrow \quad \exists\, P \in \mathcal{P} : d_P \geq \omega_P. \qquad \square$$

Nontriviality will be required to prove a necessary and sufficient condition for equilibria in games with strictly increasing costs; see Theorem 2.46.

Noncompensative Security Property

Definition 2.38. *A path player game is called a game with the* noncompensative security (NCS) *property if for all paths $P \in \mathcal{P}$ and for all flows f_{-P} with $d_P \geq \omega_P$ it holds that*

$$\exists\, f_P \geq \omega_P \quad \text{such that} \quad \tilde{b}_P(f_P) > \kappa_P.$$

In games with the NCS property, the benefit functions prevent any player P from choosing the security payment κ_P when a flow $f_P \geq \omega_P$ is possible. If a player has the possibility to earn benefit by receiving income by his "productivity" (i.e., by getting income from the cost function c_P), he should have no reason to take advantage of the security limit. The security payment shall only be used if the player has no other choice due to the strategies of his competitors (i.e., if $d_P < \omega_P$). The NCS property is an important attribute of games as it will enable the characterization of equilibria in the context of strictly increasing cost functions; see Section 2.2.5.

Note that as we assume nonnegative costs, a game where $\kappa_P < 0$ holds for all P in $\mathcal{P}$ will satisfy the NCS property. In all other cases, to identify a game with the NCS property the following definition is helpful.

Definition 2.39. *A benefit function $b_P(f)$ with $\omega_P < r$ has the* noncompensative security (NCS) *property if*

$$\kappa_P < c_P(0, \ldots, 0, \omega_P, 0, \ldots, 0) =: c_P(\mathbf{0}_{-P}, \omega_P)$$

holds.

The value of κ_P is sufficiently small such that a player on an underloaded path gets a benefit lower than the income she would get if she were able to route a flow of value ω_P over that path, if no other player routes anything. Again, the idea is that no player should have an incentive to choose her path to be underloaded while she could be able to route a flow $f_P \geq \omega_P$.

To illustrate benefit functions with the NCS property, consider a benefit function $b_P(f)$, where all players apart from P are routing a zero-flow; that is, $b_P(f) = b_P(\mathbf{0}_{-P}, f_P)$. A function $b_P(f)$ such as that shown in Figure 2.8

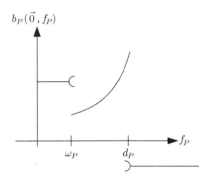
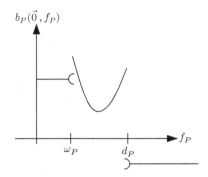

Fig. 2.8. No NCS property.   **Fig. 2.9.** NCS property.

does not satisfy the NCS property as the player P will choose the security payment instead of the income obtained by routing ω_P. In general, this does not mean that the player will always prefer the benefit κ_P. It may happen (as in this illustration) that there is a flow $f_P > \omega_P$ with $b_P(\mathbf{0}_{-P}, f_P) > \kappa_P$. But, for a benefit function without the NCS property, we cannot guarantee that there will be a flow f_P that provides a higher benefit than κ_P.

However, for a benefit function without the NCS property, we cannot guarantee that there will be a flow f_P that provides a higher benefit than κ_P. On the contrary, a benefit function such as the one shown in Figure 2.9 allows the player to obtain a benefit higher than κ_P when routing $f_P = \omega_P$.

Consider the relation between games with the NCS property and benefit functions with the NCS property. Unfortunately, a game that includes functions with the NCS property is not necessarily a game with the NCS property. Consider a path P with $d_P \geq \omega_P$, whose benefit functions satisfy the NCS property. It does not necessarily hold that P is in any case able to obtain a benefit greater than κ_P. If this player P shares an edge e with other players, and if e has a cost function with decreasing regions, it may happen that the additional flow on e, produced by the competitors, decreases the benefit of player P. So it can happen that the benefit may be smaller than the benefit P would get if he were the only one sending flow. In particular the benefit $b_P(\omega_P)$ could get smaller than or equal to κ_P for flows $f_P \geq \omega_P$; that is, the game does not satisfy the NCS property. We call this effect of influencing the benefit of the competitors the *edge-sharing effect* and analyse it in more detail in Section 2.2.4. Note that in this situation, κ_P satisfies the purpose of being a security payment, as it helps player P to escape the harmful behavior of competitors.

Nevertheless, both previous definitions are related and we investigate how benefit functions with the NCS property induce games with the NCS property. The following lemmas ensure sufficient conditions such that the NCS property of benefit functions characterizes games with the NCS property, which is important for the characterization of equilibria.

Lemma 2.40. *Consider a path player game with cost functions c_e that are monotonically increasing for all edges $e \in E$ and benefit functions $\tilde{b}_P(f_P)$ that satisfy the NCS property for all paths $P \in \mathcal{P}$. Such a game is a game with the NCS property.*

Proof. Consider a path $P \in \mathcal{P}$ and a flow f_{-P} with $d_P \geq \omega_P$. By the concept of the benefit function, for all $f_P \in [\omega_P, d_P]$ we get that

$$\tilde{b}_P\left(f_P\right) = \tilde{c}_P(f_P)$$

$$= \sum_{e \in P} c_e \left(f_P + \sum_{P_k \in \mathcal{P} \backslash \{P\} : e \in P_k} f_{P_k} \right) \tag{2.7}$$

$$\geq \sum_{e \in P} c_e \left(\omega_P + \sum_{P_k \in \mathcal{P} \backslash \{P\} : e \in P_k} 0 \right) \tag{2.8}$$

$$= c_P\left(\mathbf{0}_{-P}, \omega_P\right) > \kappa_P. \tag{2.9}$$

Condition (2.7) holds due to the definition of $c_P(f)$, and (2.8) holds because of the monotonically increasing cost functions c_e. Finally, (2.9) is true as $b_P(f)$ satisfies NCS property.

We conclude that $\tilde{b}_P(f_P) > \kappa_P$ for all $P \in \mathcal{P}$ and for all feasible f with $f_P \geq \omega_P$ and thus the game has the NCS property. □

Lemma 2.41. *Consider a path player game on a path-disjoint network G. Furthermore, let the benefit functions $b_P(f)$ satisfy the NCS property for all paths $P \in \mathcal{P}$. Such a game is a game with the NCS property.*

Proof. Consider a path $P \in \mathcal{P}$ and a flow f_{-P} with $d_P \geq \omega_P$ and set $f_P = \omega_P$. As the resulting flow f is feasible, it holds that

$$\tilde{b}_P(f_P) = \tilde{c}_P(f_P)$$
$$= c_P(\mathbf{0}_{-P}, f_P) > \kappa_P. \tag{2.10}$$

Note that (2.10) holds because G is path-disjoint (see Corollary 2.34) and as $b_P(f)$ has the NCS property. Hence, the lemma follows. □

Definition 2.42. *We denote the* set of exclusively used edges *of player P by*

$$E_P^{\text{exc}} = \{e : e \in P \land e \notin P_k \ \forall \ P_k \neq P\}.$$

The following lemma does not require benefit functions with the NCS property, but a similar condition for at least one edge in each path, to obtain a game with the NCS property.

Lemma 2.43. *A path player game where each path P satisfies that the set of exclusively used edges E_P^{exc} is nonempty, satisfies the NCS property if*

$$\sum_{e \in E_P^{\text{exc}}} c_e(\omega_P) > \kappa_P \ \forall \ P \in \mathcal{P}.$$

Proof. For a flow f consider any path P with $d_P \geq \omega_P$ and the corresponding flow f_{-P}. Set $f_P = \omega_P$; that is, the resulting flow is feasible. Then, we obtain that

$$\tilde{b}_P(f_P) = \tilde{c}_P(\omega_P)$$

$$= \overbrace{\sum_{e \in E_P^{\text{exc}}} c_e(\omega_P)}^{>\kappa_P} + \overbrace{\sum_{e \in P \setminus E_P^{\text{exc}}} c_e(f_e)}^{\geq 0} > \kappa_P,$$

and thus the lemma follows. □

The Edge-Sharing Effect

In a path player game with a general network topology paths may share edges. The cost of a common edge e is dependent on the flow f_e which can be described as the sum of flows on paths that contain e: $f_e = \sum_{P:e \in P} f_P$. The cost function c_e may contain decreasing regions, so it can happen that increasing flow decreases benefit. Here increasing flow on a path containing that edge only makes sense if this loss is compensated by increased benefits on other edges along that path. Hence it is possible that some of the players sharing edge e have incentive to raise the flow f_e even if edge e induces a loss. For instance see Figure 2.10, where P_1 would accept a decreasing income from edge e, as this loss is compensated by the remaining edges. At the same time, P_2 does not want to increase f_e too much, as at a certain point his benefit $b_P(f)$ will decrease. Nevertheless, P_2 cannot avoid that P_1 increases the flow; that is, he is forced into a situation where sending flow can create loss. Here a positive security limit together with a positive security payment can help the second player, as then her benefit need not fall below the security payment. In this case the security payment serves as protection against the harmful behavior of competitors.

The edge-sharing effect plays an important role for the NCS property of games (see Section 2.2.4). Even if all the benefit functions of the game satisfy

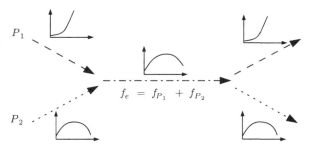

Fig. 2.10. Edge-sharing effect.

the NCS property, the edge-sharing effect can destroy the NCS property of the game. This is true as the NCS property for a path has been defined for a flow $f = (\mathbf{0}_{-P}, f_P)$. For a flow $f_{-P} \gneq \mathbf{0}_{-P}$ it is possible that there is no $f_P \geq \omega_P$ which allows $b_P(f_P) > \kappa_P$. Hence, the NCS property does not have to hold. Nevertheless, as we have already proved in Lemmas 2.40 and 2.41, the NCS property of a game may follow from benefit functions with the NCS property if additional assumptions are required which prevent the edge-sharing effect.

Clearly, the edge-sharing effect cannot take place if the observed network is path-disjoint. Also, the edge-sharing effect does not occur if we assume monotonically increasing cost functions on all edges $e \in E$ because in this case it will never happen that increasing flow causes a decrease of benefit, which is the main property of the edge-sharing effect. Finally, the harmful influence of the edge-sharing effect can be softened if each player owns at least one edge exclusively and if this exclusive edge earns "enough" income. The income of the exclusive edge will not be reduced by the edge-sharing effect. As we have assumed $c_e(f_e) \geq 0 \ \forall \ f_P \geq 0$, none of the edges can deliver negative income. This approach is used in Lemma 2.43 to ensure a game with the NCS property. In general, although the edge-sharing effect may be harmful for a player, it will never create negative costs, as the cost functions are assumed to be nonnegative for all nonnegative flow.

2.2.5 Equilibria for Special Cost Functions

In this section, we present characterizations of equilibria under special assumptions on the cost functions. If we assume strictly increasing cost functions on all edges, we obtain a necessary condition for equilibria and even a necessary and sufficient condition if the game has in addition the NCS property or if we consider a game with no security limit. Assuming differentiable costs we find a necessary condition that will also become a sufficient condition in the case of differentiable and concave functions in a game with no security limit. Finally, for convex costs we determine a dominating strategy set.

Strictly Increasing Cost Functions

We assume the cost functions c_e to be strictly increasing on all edges $e \in E$. Furthermore, to obtain some of the results, we require no security limit; that is, we set the security limit $\omega_P = 0$ for all paths $P \in \mathcal{P}$. The next proposition is useful for the proofs in this section.

Proposition 2.44. *Consider a path player game with strictly increasing cost functions c_e. Then the one-dimensional benefit functions $\tilde{b}_P(f_P)$ are also strictly increasing for $f_P \in [\omega_P, d_P]$.*

The proof of this proposition is based on the fact that $\tilde{c}_P(f_P)$ is a sum of strictly increasing functions $c_e(f_P + \sum_{P_k \in \mathcal{P} \setminus \{P\}} f_{P_k})$ and so it is again strictly increasing.

Assuming ω_P to be equal to zero, the benefit function and the one-dimensional benefit functions take the following simplified form that will appear every time we require no security limit (see page 44 for the case of differentiable and concave costs and page 46 for convex cost functions).

$$b_P(f) = \begin{cases} c_P(f) & if \ \sum_{P \in \mathcal{P}} f_P \leq r \\ -M & if \ \sum_{P \in \mathcal{P}} f_P > r \end{cases}, \qquad (2.11)$$

$$\tilde{b}_P(f_P) = \begin{cases} \tilde{c}_P(f_P) & if \ f_P \leq d_P \\ -M & if \ f_P > d_P \end{cases}. \qquad (2.12)$$

Figure 2.11 illustrates a benefit function with a strictly increasing cost function and $\omega_P = 0$. The following theorem states that in this case each flow that uses the complete flow rate r is an equilibrium and that there exist no other feasible equilibria for this kind of instance.

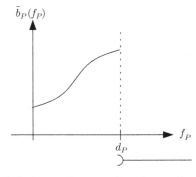

Fig. 2.11. Strictly increasing cost function and no security limit.

Theorem 2.45. *In a path player game with strictly increasing cost functions c_e on all edges $e \in E$ and security limit $\omega_P = 0$ for all $P \in \mathcal{P}$, a flow f is a feasible equilibrium if and only if*

$$\sum_{P \in \mathcal{P}} f_P = r.$$

Proof.
Part (a) f equilibrium $\Rightarrow \sum_{P \in \mathcal{P}} f_P = r$.
Let f be a feasible equilibrium and assume that $\sum_{P \in \mathcal{P}} f_P < r$.

$$\Rightarrow f_P < d_P \ \forall \ P \in \mathcal{P}$$

$$\Rightarrow \tilde{b}_P(f_P) < \tilde{b}_P(f_P + \varepsilon) \ \forall \ \varepsilon \in (0, d_P - f_P], \ \forall \ P \in \mathcal{P} \qquad (2.13)$$

$$\Rightarrow f_P \notin f_P^{\max}(f_{-P}),$$

(i.e., f is not an equilibrium) and hence,

$$\sum_{P \in \mathcal{P}} f_P = r.$$

Note that (2.13) follows because by Proposition 2.44, $\tilde{b}_P(f_P)$ is for all P in $\mathcal{P}$ and for all $f_P \in [\omega_P, d_P]$ strictly increasing in f_P.

Part (b) $\sum_{P \in \mathcal{P}} f_P = r \Rightarrow f$ equilibrium.
As $\sum_{P \in \mathcal{P}} f_P = r$ holds, it implies that $f_P = d_P$ for all $P \in \mathcal{P}$. Furthermore, for all $e \in E$, $\tilde{c}_e(f_e)$ is a strictly increasing function, thus $\tilde{b}_P(f_P)$ is strictly increasing over $[0, d_P]$ by Proposition 2.44 and $f_P \in f_P^{\max}(f_{-P}) \ \forall \ P \in \mathcal{P}$. Hence, by Corollary 2.28, f is an equilibrium. $\qquad \square$

Next, we investigate situations where we cannot assume $\omega_P = 0$ for all paths $P \in \mathcal{P}$. In this case, it could happen that a player will prefer to set her flow f_P smaller than the security limit ω_P if κ_P is higher than $\tilde{b}_P(d_P)$. Thus, the sufficient and necessary condition of Theorem 2.45 does not hold any more. Before we analyze this situation in general, we want to investigate in which cases the statement of Theorem 2.45 holds in spite of a general security limit. In fact, for strictly increasing cost functions and a general security limit it is possible to ensure the sufficient and necessary condition from Theorem 2.45 if we assume in addition having a nontrivial game with the NCS property (see Definitions 2.36 and 2.38 on pages 30 and 31).

Theorem 2.46. *Consider a game with strictly increasing cost functions c_e on all edges $e \in E$. Assume that the game is nontrivial and it satisfies the NCS property. Then a flow f is a feasible equilibrium if and only if $\sum_{P \in \mathcal{P}} f_P = r$.*

Proof.
Part (a) f is a feasible equilibrium $\Rightarrow \sum_{P \in \mathcal{P}} f_P = r$.

Consider a feasible equilibrium f and assume that $\sum_{P \in \mathcal{P}} f_P < r$, that is, $f_P < d_P$ for all $P \in \mathcal{P}$. Due to nontriviality and by Lemma 2.37 we can find a path $\bar{P}$ such that $d_{\bar{P}} \geq \omega_{\bar{P}}$.
We distinguish two cases:

(i) $f_{\bar{P}} \geq \omega_{\bar{P}}$

  $\Rightarrow \tilde{b}_{\bar{P}}(d_{\bar{P}}) > \tilde{b}_{\bar{P}}(f_{\bar{P}})$ (due to Proposition 2.44), which contradicts f being a feasible equilibrium.

(ii) $f_{\bar{P}} < \omega_{\bar{P}}$

  $\Rightarrow \exists\, \hat{f}_{\bar{P}} \geq \omega_{\bar{P}}$ such that $\tilde{b}_{\bar{P}}(\hat{f}_{\bar{P}}) > \kappa_{\bar{P}} = \tilde{b}_{\bar{P}}(f_{\bar{P}})$ (due to the NCS property), which contradicts f being a feasible equilibrium.

The above implies that $\sum_{P \in \mathcal{P}} f_P = r$.

Part (b) $\sum_{P \in \mathcal{P}} f_P = r \Rightarrow f$ is a feasible equilibrium.
Consider a flow with $\sum_{P \in \mathcal{P}} f_P = r$, that is, $f_P = d_P$ for all $P \in \mathcal{P}$.
Consider the two cases:

(i) $f_P \geq \omega_P$

  As there exists at least one $\hat{f}_P \geq \omega_P$ such that $\tilde{b}_P(\hat{f}_P) > \kappa_P$ (due to the NCS property), and as $\tilde{b}_P(f_P)$ is strictly increasing over $[\omega_P, d_P]$ (see Proposition 2.44), and in particular, $\tilde{b}_P(f_P) \geq \tilde{b}_P(\hat{f}_P) > \kappa_P$, it holds that $f_P^{\max} = \{d_P\}$.

(ii) $f_P < \omega_P$

  As $\tilde{b}_P(f_P)$ is constant over $[0, \omega_P)$ and $d_P < \omega_P$ it holds that $d_P \in f_P^{\max}$.

We conclude that f is a feasible equilibrium due to $f_P \in f_P^{\max} \,\forall\, P \in \mathcal{P}$.   □

Unfortunately, the reverse of Theorem 2.46 does not hold. A game that satisfies the property:

"A flow f is a feasible equilibrium if and only if $\sum_{P \in \mathcal{P}} f_P = r$"   (2.14)

does not have to be nontrivial nor satisfy the NCS property. For an illustration we present the following examples.

Example 2.47. (2.14) $\nRightarrow$ NCS property.
Consider a game on a network with two paths, as illustrated in Figure 2.12. The flow rate is given by $r = 1$. On both paths the costs are $c_P(x) = x$, but the security limits and security payments differ: $\omega_1 = \kappa_1 = 1$ and $\omega_2 = \kappa_2 = 0$. In this game a flow f with $\sum_{P \in \mathcal{P}} f_P < r$ cannot be an equilibrium due to $f_2^{\max} = \{d_2\}$ for all f_{-2}. That means player 2 would in any case use up the remaining flow rate. On the other hand, each flow f with $\sum_{P \in \mathcal{P}} f_P = r$ is an equilibrium. If $\sum_{P \in \mathcal{P}} f_P = r$ holds, player 2 cannot find any better strategy as he will always try to get as much flow as possible, whereas player 1 is also not able to improve her payoff as her benefit function is anyway constant over

$[0, 1]$. That means this game satisfies condition (2.14). Nevertheless, the game has no NCS property. There is no $f_1 \geq \omega_1$ with $\tilde{b}_1(f_1) > \kappa_1$ and so player 1 destroys the NCS property of the game.

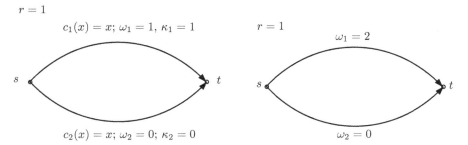

Fig. 2.12. Network of Example 2.47.   **Fig. 2.13.** Network of Example 2.48.

Example 2.48. $(2.14) \not\Rightarrow$ nontriviality.
Consider the game illustrated in Figure 2.13. The graph consists of two paths, and we choose $\omega_1 = 2$ and $\omega_2 = 0$. The remaining components of the game, such as cost functions and security payments, may be chosen arbitrarily, but it is important that the cost functions are strictly increasing. With a similar argument as in Example 2.47, it is possible to show that this game satisfies condition (2.14). Nevertheless, the game is trivial, because $\sum_{P \in \mathcal{P}} \omega_P > r$.

If we consider a game with strictly increasing cost functions and a general security limit, but cannot ensure the NCS property or the nontriviality of the game, we are still able to provide information about equilibria for this kind of instance.

Lemma 2.49. *If a flow f in a path player game with strictly increasing cost functions c_e on all edges $e \in E$ is a feasible equilibrium then at least one of the following two cases holds.*

(i) $\sum_{P \in \mathcal{P}} f_P = r$.
(ii) $f_P < \omega_P \; \forall \; P \in \mathcal{P}$.

Proof. Let f be a feasible equilibrium and assume that (i) is not true; that is, $\sum_{P \in \mathcal{P}} f_P < r$. Then $f_P < d_P \; \forall \; P \in \mathcal{P}$.
Assume case (ii) is also false; that is, $\exists \; \bar{P}$ with $f_{\bar{P}} \geq \omega_{\bar{P}}$. Then $\tilde{b}_{\bar{P}}(f'_{\bar{P}}) > \tilde{b}_{\bar{P}}(f_{\bar{P}}) \; \forall \; f'_{\bar{P}} \in (f_{\bar{P}}, d_{\bar{P}}]$, as according to Proposition 2.44, $\tilde{b}_P(f_P)$ is strictly increasing over this domain. It follows that $f_{\bar{P}} \notin f_{\bar{P}}^{\max}$. This contradicts f being an equilibrium, hence $f_P < \omega_P$ for all $P \in \mathcal{P}$. $\qquad \square$

To illustrate Lemma 2.49 we present two examples of feasible equilibria where (i) and (ii) do not hold.

Example 2.50. Consider a path player game with two vertices s and t that are connected by two edges, $\mathcal{P} = \{1, 2\}$. A flow rate $r = 1$ has to be routed between the two vertices. We set the security limits $\omega_1 = \omega_2 = 0.5$, the security payment $\kappa_1 = \kappa_2 = 1$, and the cost functions $c_P(x) = x$, $P \in \{1, 2\}$. The flow $f = (0.2, 0.2)$ with $b_1(f) = b_2(f) = 1$ is an equilibrium for which (ii) holds and (i) is not satisfied.

Example 2.51. Consider the following path player game. There are two vertices s and t connected by two edges, $\mathcal{P} = \{1, 2\}$. A flow rate $r = 2$ has to be routed from s to t. Furthermore, the paths have security limits $\omega_P = 1$, security payments $\kappa_P = 1$, and cost functions $c_P(f_P) = 1 + f_P$.
The flow $f = (0.5, 1.5)$ with $\tilde{b}_1(0.5) = 1$ and $\tilde{b}_2(1.5) = 2.5$ destroys property (ii) and (i) does hold. This flow is an equilibrium as none of the players is able to improve the current payoff.

The following lemma provides a statement about the reverse of Lemma 2.49.

Lemma 2.52. *Consider a path player game with strictly increasing cost functions c_e. Let f be a flow with the following properties.*

(i) $\sum_{P \in \mathcal{P}} f_P = r$.
(ii) $f_P < \omega_P \; \forall \; P \in \mathcal{P}$.

Then, f is a feasible equilibrium.

Proof. For all players P in $\mathcal{P}$ and for all $\varepsilon > 0$, we have:

$$\tilde{b}_P(f_P + \varepsilon) = -M < \tilde{b}_P(f_P),$$

$$\tilde{b}_P(f_P - \varepsilon) = \kappa_P = \tilde{b}_P(f_P), \text{ if } \varepsilon \leq f_P.$$

It follows that for all $P \in \mathcal{P}$ and for all $\bar{f}_P > 0$,

$$\tilde{b}_P(f_P) \geq \tilde{b}_P(\bar{f}_P)$$

holds, and hence f is a feasible equilibrium. $\qquad\square$

The following two examples illustrate that a flow that satisfies only one of the conditions (i) and (ii) does not have to be an equilibrium.

Example 2.53. $((i) \land \lnot(ii) \nRightarrow f$ is feasible equilibrium)
Consider a path player game with strictly increasing cost functions c_e. Furthermore, consider a feasible flow f such that $\sum_{P \in \mathcal{P}} f_P = r$ holds and that there exists $\bar{P} \in \mathcal{P}$ with $f_{\bar{P}} \geq \omega_{\bar{P}}$. It is possible to construct a game such that $\tilde{b}_{\bar{P}}(f_{\bar{P}}) < \kappa_{\bar{P}}$ holds (see Figure 2.14) and thus, the flow f is not an equilibrium.
Set $r = 1$, $\omega_1 = \omega_2 = 0.25$, and the security payment $\kappa_1 = \kappa_2 = 2$. For cost functions $c_P(x) = x$ with $P = \{1, 2\}$ the flow $f = (0.5, 0.5)$ satisfies (i) but not (ii). This flow with $b_1(f) = b_2(f) = 0.5$ is not an equilibrium because $f_1^{\max} = f_2^{\max} = [0, 0.25)$.

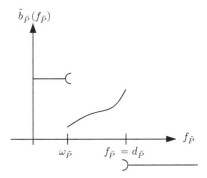

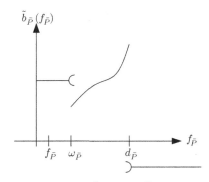

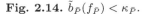

Fig. 2.14. $\tilde{b}_{\bar{P}}(f_{\bar{P}}) < \kappa_{\bar{P}}.$ **Fig. 2.15.** $\tilde{b}_{\bar{P}}(f_{\bar{P}}) < \tilde{b}_{\bar{P}}(d_{\bar{P}}).$

Example 2.54. ($\neg(i) \wedge (ii) \not\Rightarrow f$ is feasible equilibrium)
Consider a path player game with strictly increasing cost functions c_e. Furthermore, consider a feasible flow f such that $\sum_{P \in \mathcal{P}} f_P < r$ and $f_{\bar{P}} < \omega_{\bar{P}}$
holds for all $P \in \mathcal{P}$. Thus, it holds for all P that $f_P < d_P$ and it is possible
to construct a game such that

$$\exists \, \bar{P} : \tilde{b}_{\bar{P}}(f_{\bar{P}}) = \kappa_{\bar{P}} < \tilde{b}_{\bar{P}}(d_{\bar{P}})$$

holds (see Figure 2.15); that is, f is not an equilibrium.
  Set $r = 1$, $\omega_1 = \omega_2 = 0.5$, and $\kappa_1 = \kappa_2 = 0.1$. For cost functions $c_P(x) = x$, $P = \{1, 2\}$ a flow $f = (0.45, 0.45)$ with $b_1(f) = b_2(f) = 0.1$ is no equilibrium
due to $f_1^{\max} = f_2^{\max} = 0.55$.

  We have found out that a feasible flow with property (ii) need not be an
equilibrium. This doesn't change if we assume having a trivial game or a game
with no NCS property.

Remark 2.55. Consider a path player game with strictly increasing cost functions c_e. Furthermore, consider a feasible flow f, such that

$$f_P < \omega_P \,\, \forall \, P \in \mathcal{P} \qquad (ii)$$

holds. This flow is not necessarily an equilibrium even if the game is trivial
or if it does not satisfy the NCS property or both.

  We present an example for proving Remark 2.55.

Example 2.56. Consider a game with two vertices s and t and two edges connecting the vertices ($\mathcal{P} = \{1, 2\}$ on a path-disjoint network). A flow rate $r = 5$
has to be routed from s to t. Both paths have a security limit $\omega_P = 3$; that is,
the game is trivial. Furthermore, the security payment is $\kappa_P = 1$ for $P \in \{1, 2\}$
and the cost functions are $c_1(f_1) = f_1$ and $c_2(f_2) = f_2/10$. This game does not
satisfy the NCS property because there is no $f_2 > \omega_2$ such that $\tilde{b}_2(f_2) > \kappa_2$.

Consider the feasible flow $f = \mathbf{0}_{|\mathcal{P}|}$. The flow f satisfies (ii) and $d_P = r$ for all P. Nevertheless, because $\tilde{b}_1(d_1) = 3 > \tilde{b}_1(0) = \kappa_1 = 1$ this flow is not an equilibrium.

Differentiable Cost Functions

Let the cost functions c_e be differentiable over $(0, r)$ for all edges $e \in E$. We need to define a quasi-derivative of the one-dimensional benefit function $\tilde{b}_P(f_P)$ over the domain $[0, d_P]$. Because a security limit $\omega_P > 0$ may cause a jump at $f_P = \omega_P$ we have to take care of that in the definition.

Definition 2.57. *Consider a path player game with cost functions that are differentiable over $(0, r)$ and a given flow f_{-P} with $d_P > 0$. The quasi-derivative of the one-dimensional benefit function $\tilde{b}_P(f_P)$ for all $f_P \in [0, d_P]$, is denoted by*

$$
\tilde{b}'_P(f_P) = \begin{cases} 0 & \text{if } f_P < \omega_P \\ \tilde{c}'_P(f_P+) & \text{if } f_P = \omega_P \\ \tilde{c}'_P(f_P-) & \text{if } f_P = d_P \\ \tilde{c}'_P(f_P) & \text{otherwise} \end{cases}
$$

with

$$
\tilde{c}'_P(f_P+) = \lim_{\varepsilon \to 0, \varepsilon > 0} \frac{\tilde{c}_P(f_P + \varepsilon) - \tilde{c}_P(f_P)}{\varepsilon},
$$

$$
\tilde{c}'_P(f_P-) = \lim_{\varepsilon \to 0, \varepsilon < 0} \frac{\tilde{c}_P(f_P + \varepsilon) - \tilde{c}_P(f_P)}{\varepsilon},
$$

$$
\tilde{c}'_P(f_P) = \frac{\partial \left[\sum_{e \in P} c_e(f_P + \sum_{P_k : e \in P_k \wedge P_k \neq P} f_{P_k}) \right]}{\partial f_P}.
$$

As the one-dimensional benefit function $\tilde{b}_P(f_P)$ may not be differentiable over $f_P \in [0, d_P]$, it is not formally correct to call the construction of Definition 2.57 a "derivative". Nevertheless, in our context we use the term "quasi-derivative", as it describes very well the nature of that construction.

Lemma 2.58. *Let f be a feasible equilibrium in a path player game with differentiable cost functions c_e for all $e \in E$. Then for all $P \in \mathcal{P}$ one of the following three statements holds true.*

(i) $\tilde{b}'_P(f_P) > 0 \ \wedge \ f_P = d_P.$
(ii) $\tilde{b}'_P(f_P) < 0 \ \wedge \ f_P = \omega_P.$
(iii) $\tilde{b}'_P(f_P) = 0.$

Proof.

Part (a) (i).

Consider a path P with $\tilde{b}'_P(f_P) > 0$ and assume that (i) does not hold; that is, $f_P < d_P$. As $\tilde{b}'_P(f_P) > 0$ there is an $\varepsilon \in (0, d_P - f_P]$ such that $\tilde{b}_P(f_P + \varepsilon) > \tilde{b}_P(f_P)$; that is, $f_P \notin f_P^{\max}(f_{-P})$. This contradicts the fact that f is an equilibrium and because f is feasible, we can conclude that $f_P = d_P$.

Part (b) (ii).

Consider a path P with $\tilde{b}'_P(f_P) < 0$ and assume that (ii) does not hold; that is, $f_P > \omega_P$. As $\tilde{b}_P(f_P) < 0$ an $\varepsilon \in (0, f_P - \omega_P]$ can be found such that $\tilde{b}_P(f_P - \varepsilon) > \tilde{b}_P(f_P)$; that is, $f_P \notin f_P^{\max}(f_{-P})$. This contradicts f being an equilibrium and we can conclude that f_P has to equal zero.

Part (c) (iii).

If (i) and (ii) do not hold then (iii) has to hold. □

Figures 2.16 to 2.18 illustrate Lemma 2.58 for an arbitrary path P. Each picture shows the one-dimensional benefit function $\tilde{b}_P(f_P)$ of P. In Figure 2.16 the quasi-derivative $\tilde{b}'_P(f_P)$ of the one-dimensional benefit function is positive for the flow $f_P = d_P$. Figure 2.17 corresponds to case (ii). Here $\tilde{b}'_P(d_P) < 0$ holds, and f_P equals zero. If $\tilde{b}'_P(f_P) = 0$ for a feasible flow f, then f_P may lie anywhere in $[0, d_P]$. Figure 2.18 illustrates a flow $f_P \in (0, d_P)$.

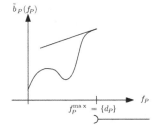

Fig. 2.16. $\tilde{b}'_P(f_P) > 0$.

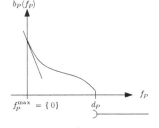

Fig. 2.17. $\tilde{b}'_P(f_P) < 0$.

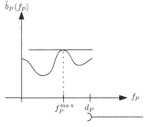

Fig. 2.18. $\tilde{b}'_P(f_P) = 0$.

The next corollary follows directly from Lemma 2.58.

Corollary 2.59. *Let f be a feasible equilibrium in a path player game with differentiable cost functions c_e for all $e \in E$. It holds that*

$$f_P > \omega_P \ \Rightarrow \ \tilde{b}'_P(f_P) \geq 0.$$

Example 2.60. Consider a path consisting of a single edge that is not shared with any other path; that is, $\tilde{c}_P(f_P) = c_e(f_e)$, and choose $c_e(f_e) = 2f_e - f_e^2$. See Figure 2.19 for an illustration of that function, where $d_P = 2$. We choose

$\omega_P = 0$. The quasi-derivative of the one-dimensional benefit function is given by $\tilde{b}'_P(f_P) = 2 - 2f_P$; that is, $\tilde{b}'_P(f_P)$ is nonnegative for $f_P \in [0, 1]$. Hence, a flow f with $f_P > 1$ will never be an equilibrium.

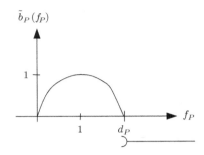

Fig. 2.19. Illustration of Example 2.60.

Differentiable and Concave Cost Functions

We now assume the cost functions c_e to be differentiable and concave for all $e \in E$. From Lemma 2.18 (see pag 20) we know that in this case also $\tilde{c}_P(f_P)$ are concave functions for all $P \in \mathcal{P}$. This case is a particular instance of path player games with differentiable cost functions described in the previous paragraph and thus, Lemma 2.58 holds. If we consider games with no security limit in addition we are in fact able to present a similar and even stronger "if and only if" result.

We set the security limit to zero: $\omega_P = 0$ for all $P \in \mathcal{P}$. Due to no security limit, the benefit function and the one-dimensional benefit functions take the form given in (2.11) and (2.12) on page 36. We use the notation of the quasi-derivative $\tilde{b}'_P(f_P)$ of the one-dimensional benefit function (see Definition 2.57). The following lemma corresponds to Lemma 2.58 but is adjusted to a no security limit game.

Theorem 2.61. *Consider a flow f in a path player game with differentiable and concave cost functions c_e for all $e \in E$ and $\omega_P = 0$ for all $P \in \mathcal{P}$. The flow f is a feasible equilibrium if and only if for all $P \in \mathcal{P}$ one of the following three cases is satisfied.*

(i) $\tilde{b}'_P(f_P) > 0 \;\wedge\; f_P = d_P.$
(ii) $\tilde{b}'_P(f_P) < 0 \;\wedge\; f_P = 0.$
(iii) $\tilde{b}'_P(f_P) = 0.$

For the proof, we need the following lemma, which proof follows from the mean value theorem.

Lemma 2.62. *Consider a function g that is continuous over $[p, g]$, differentiable and concave over (p, q), and let g' denote its derivative. Then*

(i) $g'(p) < 0$ ⇒ p *maximizes g over $[p, q]$, and*
(ii) $g'(q) > 0$ ⇒ q *maximizes g over $[p, q]$*

holds.

Proof.
Part (a) (i).

Assume that $g'(p) < 0$ holds. As g is differentiable and concave over (p, q), it holds that g' is monotonically decreasing over that interval (see Section 3.1.5.4 in [BS91]). That means $g'(\xi) \leq 0$ for all $\xi \in (p, q)$. Due to the mean value theorem there exists a $\xi \in (p, x)$ such that $g(x) = g(p) + g'(\xi)(x - p)$ holds. Thus, $g(x) \leq g(p)$ for all $x \in [p, q]$ and the proposition follows.

Part (b) (ii).

Can be proved analogously to part (a). □

In the following, we present the proof of Theorem 2.61.

Proof (Proof of Theorem 2.61).
Part (a) f feasible equilibrium ⇒ *(i) or (ii) or (iii).*

Follows from Lemma 2.58.

Part (b) *(i) or (ii) or (iii)* ⇒ f equilibrium.

(i) Let $\tilde{b}'_P(f_P) > 0$ and $f_P = d_P$. Due to Lemma 2.62, the flow f_P maximizes $\tilde{b}_P(f_P)$ over $[0, d_P]$ and hence $f_P \in f_P^{\max}$.
(ii) Let $\tilde{b}'_P(f_P) < 0$ and $f_P = 0$. Due to Lemma 2.62, the flow f_P maximizes $\tilde{b}_P(f_P)$ over $[0, d_P]$ and hence $f_P \in f_P^{\max}$.
(iii) Consider a path P with $\tilde{b}'_P(f_P) = 0$. As $\tilde{b}'_P(f_P) = \tilde{c}'_P(f_P)$ for $f_P \leq d_P$ and $\tilde{c}_P$ is a concave function it follows that f_P is the global maximum of $\tilde{b}(f_P)$ (see Theorem 3.4.2 in [BS79]). We can conclude that $f_P \in f_P^{\max}(f_{-P})$. □

Figures 2.20 to 2.22 illustrate the cases (i)–(iii). For the last case, in which $\tilde{b}'_P(f_P) = 0$ holds, the maximum in other instances may also be attained at zero or d_P. After considering path player games with differentiable and concave cost functions and no security limit, we next let the security limit ω_P be any arbitrary value.

As mentioned before, as the considered game is a special instance of a game with differentiable cost functions (see Section 2.2.5), Lemma 2.58 holds. That means that for a feasible equilibrium f each path P in $\mathcal{P}$ satisfies one of the three conditions of Lemma 2.58.

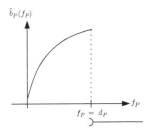

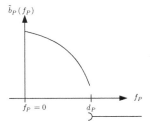

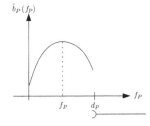

Fig. 2.20. $\tilde{b}'_P(f_P) > 0.$ **Fig. 2.21.** $\tilde{b}'_P(f_P) < 0.$ **Fig. 2.22.** $\tilde{b}'_P(f_P) = 0.$

(i) $\tilde{b}'_P(f_P) > 0 \ \wedge \ f_P = d_P.$
(ii) $\tilde{b}'_P(f_P) < 0 \ \wedge \ f_P = \omega_P.$
(iii) $\tilde{b}'_P(f_P) = 0.$

For games with no security limit, we have been able to prove also the converse direction; that is, if one of the three cases holds for each P, the considered flow is a feasible equilibrium. Unfortunately, this is not possible for general security limits. For instance see Figure 2.23. If we set $f_P = \omega_P$, case (ii) holds: $\tilde{b}'_P(f_P) < 0$ and $f_P = \omega_P$. Nevertheless, the corresponding flow f is not a feasible equilibrium, due to $f_P^{\max} = [0, \omega_P)$.

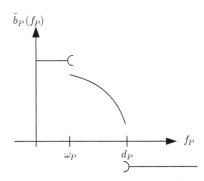

Fig. 2.23. $f_P = \omega_P$ satisfies (ii).

Convex Cost Functions

We analyze the path player game with cost functions c_e which are convex for all $e \in E$. Note that we do not require differentiability in this section. From Lemma 2.18 we know that for convex cost functions, the one-dimensional benefit functions $\tilde{b}_P(f_P)$ are also convex in $[\omega_P, d_P]$ for all $P \in \mathcal{P}$. Furthermore, we set the security limit to zero: $\omega_P = 0 \ \forall \ P \in \mathcal{P}$.

The benefit function and the one-dimensional benefit function are given by (2.11) and (2.12) on page 36.

$$b_P(f) = \begin{cases} c_P(f) & if \ \sum_{P \in \mathcal{P}} f_P \leq r \\ -M & if \ \sum_{P \in \mathcal{P}} f_P > r \end{cases},$$

$$\tilde{b}_P(f_P) = \begin{cases} \tilde{c}_P(f_P) & if \ f_P \leq d_P \\ -M & if \ f_P > d_P \end{cases}.$$

The one-dimensional benefit function consists of a convex region created by $\tilde{c}_P(f_P)$ and a constant region created by the infeasibility penalty value $-M$. The next theorem presents the following result: to check whether a flow f is an equilibrium, it is sufficient to analyze only the extreme points of the decision intervals $[0, d_P]$ (see Definition 2.20) for all paths P.

Theorem 2.63. *In a path player game with convex cost functions c_e on all edges $e \in E$ and a security limit $\omega_P = 0$ for all paths $P \in \mathcal{P}$, a feasible flow f^* is a feasible equilibrium if and only if all paths P in $\mathcal{P}$ satisfy*

$$\tilde{b}_P(f_P^*) \geq \tilde{b}_P(f_P) \ \forall \ f_P \in \{0; d_P\}. \tag{2.15}$$

Proof.
Part (a) f is a feasible equilibrium $\Rightarrow$ (2.15).

Let f^* be a feasible equilibrium. It follows directly from Definition 2.26 that for all P in $\mathcal{P}$ it holds that

$$\tilde{b}_P(f_P^*) \geq \tilde{b}_P(f_P) \ \forall \ f_P \in [0, d_P].$$

Hence, $\tilde{b}_P(f_P^*) \geq \tilde{b}_P(f_P) \ \forall \ f_P \in \{0; d_P\}$.

Part (b) (2.15) $\Rightarrow$ f is a feasible equilibrium.

Let all P in $\mathcal{P}$ satisfy that $\tilde{b}_P(f_P^*) \geq \tilde{b}_P(f_P) \ \forall \ f_P \in \{0; d_P\}$.

As $\tilde{c}_P(f_P)$ is convex and $\tilde{b}_P(f_P) = \tilde{c}_P(f_P)$ for $f_P \leq d_P$, we can conclude that a maximum of $\tilde{b}_P(f_P)$ is attained at least at one extreme point ($f_P = 0$ or $f_P = d_P$) of the decision interval (see [BS79], Theorem 3.4.6). As

$$\tilde{b}_P(f_P^*) \geq \tilde{b}_P(f_P) \quad for \ f_P \in \{0; d_P\}$$

$$\Rightarrow \tilde{b}_P(f_P^*) \geq \max_{0 \leq f_P \leq d_P} \tilde{b}_P(f_P)$$

$$\Rightarrow f_P^* \in f_P^{\max}(f_{-P}^*). \qquad \square$$

By Theorem 2.63, there exist equilibria in the described game such that the flows are chosen among the extreme points of the decision interval. If a cost function is convex but not strictly convex, a constant array of that function may enable equilibria with $f_P \notin \{0, d_P\}$ and $\tilde{b}_P(f_P) = \max\{\tilde{b}_P(0); \tilde{b}_P(d_P)\}$. In these cases the function $\tilde{b}_P(f_P)$ is constant for $f_P \in [0, d_P]$; the next result verifies this statement.

Lemma 2.64. *Consider a path player game with convex cost functions c_e and a feasible equilibrium f^*. If there is a P such that $f_P^* \notin \{0; d_P\}$, then the function $b_P(f_P)$ is constant for $f_P \in [0, d_P]$.*

Proof. Consider the function $\tilde{b}_P(f_P)$ being convex over $[0, d_P]$ and the flow $f_P^* \in (0, d_P)$. As f^* is a feasible equilibrium it holds that $\tilde{b}_P(f_P^*) = \max\left\{\tilde{b}_P(0); \tilde{b}_P(d_P)\right\}$. Without loss of generality let it hold that $\tilde{b}_P(0) \geq \tilde{b}_P(d_P)$ and thus $\tilde{b}_P(f_P^*) = \tilde{b}_P(0)$. For $f_P \in (f_P^*, d_P]$ choose λ such that

$$f_P^* = \lambda 0 + (1 - \lambda) f_P$$

$$\Rightarrow \tilde{b}_P(f_P^*) \leq \lambda \tilde{b}_P(0) + (1 - \lambda) \tilde{b}_P(f_P).$$

Suppose $\tilde{b}_P(f_P) < \tilde{b}_P(f_P^*)$,

$$\Rightarrow \tilde{b}_P(f_P^*) \lneqq \lambda \tilde{b}_P(f_P^*) + (1 - \lambda) \tilde{b}_P(f_P^*) = \tilde{b}_P(f_P^*),$$

which is a contradiction and thus it follows that there is no $f_P > f_P^*$ such that $\tilde{b}_P(f_P) < \tilde{b}_P(f_P^*)$. As $f_P^* \in f_P^{\max}$, it holds that $\tilde{b}_P(f_P) = \tilde{b}_P(f_P^*)$ for all $f_P > f_P^*$, in particular $\tilde{b}_P(d_P) = \tilde{b}_P(f_P^*)$. As $\tilde{b}_P(f_P^*) = \tilde{b}_P(0)$ holds, all flows $f_P < f_P^*$ satisfy $\tilde{b}_P(f_P) = \tilde{b}_P(f_P^*)$ and the claim follows.   □

If we assume in addition the cost functions to be strictly convex, we can improve the result stating that all equilibria will take place at the extreme points of the decision intervals:

Lemma 2.65. *If f is a feasible equilibrium in a path player game with strictly convex cost functions c_e on all edges $e \in E$ and a security limit $\omega_P = 0$ for all paths $P \in \mathcal{P}$ then for all P in $\mathcal{P}$,*

$$f_P \in \{0; d_P\}.$$

Proof. Let f be a feasible equilibrium. As $\tilde{b}_P(f_P) = \tilde{c}_P(f_P)$ for $f_P \leq d_P$ and $\tilde{c}_P(f_P)$ is strictly convex it follows that $f_P^{\max} \subseteq \{0, d_P\}$. As f is a flow, $f_P \in f_P^{\max}(f_{-P}) \subseteq \{0; d_P\}$.   □

Consider Figures 2.24 and 2.25. The left one illustrates a strictly convex cost function. In this case, an equilibrium will always provide a flow $f_P \in \{0, d_P\}$. In a convex but not strictly convex cost function it is possible that an equilibrium creates a flow $f_P \notin \{0, d_P\}$, such as illustrated in Figure 2.25. In fact, this will only happen in functions that are constant over $[0, d_P]$, as Lemma 2.64 has already shown.

We have stated in Corollary 2.21 that if there exists a P with $f_P = d_P$ then this equation holds for all P in $\mathcal{P}$. Using that observation we present the following corollary.

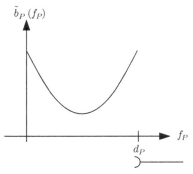

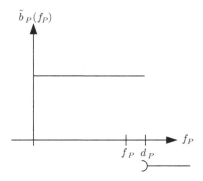

Fig. 2.24. Strictly convex costs.   **Fig. 2.25.** Convex costs.

Corollary 2.66. *Let f^* be a feasible equilibrium in a path player game with strictly convex cost functions c_e on all edges $e \in E$ and a security limit $\omega_P = 0$ for all paths $P \in \mathcal{P}$. If there is a path P with $f_P^* = 0$ and $d_P > 0$ then $f^* = \mathbf{0}_{|\mathcal{P}|}$.*

Proof. Assume $\exists \bar{P} \in \mathcal{P} : f_{\bar{P}}^* = 0 \ \wedge \ d_{\bar{P}} > 0$; that is $d_{\bar{P}} \neq f_{\bar{P}}^*$

$$\Rightarrow \nexists P \in \mathcal{P} : f_P = d_P \tag{2.16}$$

$$\Rightarrow f_P < d_P \ \forall \ P \in \mathcal{P} \tag{2.17}$$

$$\Rightarrow f_P = 0 \ \forall \ P \in \mathcal{P}. \tag{2.18}$$

Statement (2.16) follows from Corollary 2.21, and (2.17) holds as f^* is feasible. Finally, (2.18) follows from Lemma 2.65. $\qquad \square$

The corollary describes a property of path player games with strictly convex cost functions. If there is only one player who wants to set a positive flow, the one-dimensional benefit function of this player must be of a shape that allows the player to reach a region where the function is monotonically increasing; that is, right-handed from a local minimum. If he could not reach this region, he would have incentive to set as little flow as possible; for instance a zero-flow. As he is able to reach this region, he has incentive to propose as much flow as possible; for instance he will set $f_P = d_P$ and so $f_P = d_P \ \forall \ P \in \mathcal{P}$ holds.

2.3 Dominated Equilibria

2.3.1 Introduction

In path player games there may exist multiple equilibria. For instance in a game with more than one player, with strictly increasing cost functions, positive flow rate, and no security limit (see Section 2.2.5), we have an infinite number of equilibria. Each flow f^* that satisfies $\sum_{P \in \mathcal{P}} f_P^* = r$ is an equilibrium in such a game, see Theorem 2.45 on page 37. Although being stable situations, equilibria may be disadvantageous for some or even all players. For instance even an infeasible flow, where every player gets punished with a benefit of $-M$, may be an equilibrium. This fact meets the question of dominance among equilibria and flows in general. When do we, considering the interests of all players, prefer one flow more than another? We surely prefer a flow f rather than $\hat{f}$ if the benefit of f is for all players not lower and for at least one player higher than the benefit of $\hat{f}$. This observation is summarized in the definition of dominance among flows, where we use the following convention to compare vectors.

Definition 2.67. *Let $u = (u_1, \ldots, u_k)$ and $v = (v_1, \ldots, v_k)$ be k-dimensional vectors. We write $u \gneq v$ if the following holds:*

$$u_i \geq v_i \ \forall \ i = 1, \ldots, k \quad \wedge \quad \exists \ j : u_j > v_j.$$

Furthermore, we denote the *vector of benefits* by $b(f) = (b_P(f))_{P \in \mathcal{P}}$.

Definition 2.68. *A flow $\hat{f}$ is called* Pareto-dominated *(in short: dominated) if there exists a* Pareto-dominating *flow (in short: dominating flow); that is, a flow f such that*

$$b(f) \gneq b(\hat{f}).$$

Otherwise, $\hat{f}$ is called non-Pareto-dominated *(in short: nondominated).*

Note that an infeasible flow cannot dominate any other flow and is dominated by each feasible flow. Hence, we only consider feasible flows when we are referring to dominated or nondominated flows. This definition of dominance is closely related to the definition of dominance in multicriteria optimization; see, for instance, [Ehr05].

Harsanyi and Selten propose in [HS88] a different definition of dominance among flows. They call a flow $\hat{f}$ *payoff-dominated* if there is a flow f such that

$$b_P(f) > b_P(\hat{f}) \ \forall \ P \in \mathcal{P}.$$

This definition is stronger than the Pareto-dominance, as a flow is only considered to be preferable against a second flow if there is an improvement obtained for each player. Each payoff-dominated flow is also Pareto-dominated.

Another use of the term "dominance" is common in game theory. Consider the strategy set of a single player i. A strategy x dominates a second strategy y, if for all strategy choices s_{-i} of the opponents it holds for the payoff p_i of player i: $p_i(s_{-i}, x) \geq p_i(s_{-i}, y)$ and there is one $\hat{s}_{-i}$ such that $p_i(\hat{s}_{-i}, x) > p_i(\hat{s}_{-i}, y)$ holds (see, e.g., [Owe95]). Dominated strategies can be neglected by the player and be removed from the strategy set. Thus, this approach enables the reduction of the problem size and can be utilized as preprocessing in terms of computation of equilibria. This definition considers the dominance among strategies of a single player and should not be mixed up with dominance among flows, which we are using in this text.

The intention of this material is to compare equilibria and to look for nondominated ones. We present classes of path player games, where each nondominated flow is also an equilibrium and one class where also the reverse is true. Furthermore, we present one class of path player games, where equality of the sets of equilibria and nondominated flows does hold. Also, we show that disadvantageous situations related to the Prisoner's Dilemma, where each equilibrium is dominated may occur in path player games. Parts of this chapter have been published in [SS06a]. The following example illustrates the definitions.

Example 2.69. Consider a network consisting of two edges that link the nodes s and t as illustrated in Figure 2.26. The game is a game with no security limit as $\omega_1 = \omega_2 = 0$. Let the flow rate $r = 1$ and the cost functions be $c_1(x) = x$ and $c_2(x) = 1$.

The flow $f^* = (0.5, 0.5)$ with $b(f^*) = (0.5, 1)$ is a feasible equilibrium because $f_1^{\max} = \{0.5\}$ and $f_2^{\max} = [0, 0.5]$. This flow is dominated, for example, by the flows $f^{**} = (0.75, 0.25)$ and $f = (0.7, 0.25)$ due to $b(f^{**}) = (0.75, 1) \gneqq b(f^*)$ and $b(f) = (0.7, 1) \gneqq b(f^*)$. Note that f^{**} is an equilibrium itself, whereas f is not. On the other hand, consider the flow $\bar{f} = (1, 0)$ with $b(\bar{f}) = (1, 1)$. The flow $\bar{f}$ is an equilibrium that is nondominated.

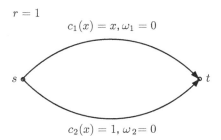

$r = 1$

$c_1(x) = x, \omega_1 = 0$

s t

$c_2(x) = 1, \omega_2 = 0$

Fig. 2.26. Game network Example 2.69.

This example illustrates that in path player games, there may exist dominated equilibria that can be improved to situations that are more preferable for the overall system. That means some players could obtain increased payoff and none of the players would have to accept a worse situation. However, these improved situations need not be equilibria and may hence be unstable. A famous example of a situation where the (unique) equilibrium respresents a disadvantageous situation for the players is the Prisoner's Dilemma (see, e.g., [FT91]). In this two-person game, there is a game situation that improves the benefit of both players compared to the unique equilibrium of the game, however, this situation is no equilibrium itself and hence unstable. To obtain a situation where the interest of all players is taken into account, we are interested in determining equilibria that are not dominated. Dominated equilibria are unsatisfying as they are stable situations; that is, the game will get stuck, although there exist better game situations.

With this observation the following question appears. What is the relation between the equilibria and the nondominated flows of a game? To analyze this question, let us denote for a game Γ, the set of equilibria with $NE(\Gamma)$ (Nash equilibria) and the set of nondominated flows with $ND(\Gamma)$. In the sequel it is demonstrated that $ND(\Gamma)$ may be contained in $NE(\Gamma)$ or vice versa that equality of the two sets may hold that the sets may be disjoint, or that neither of these possibilities may be true.

2.3.2 Relations Between Equilibria and Nondominated Flows

In this section, we show that in path player games, all possible relations between $ND(\Gamma)$ and $NE(\Gamma)$ may take place. We introduce two classes of path player games satisfying that each nondominated flow is an equilibrium. Furthermore, we provide one class where equality of these two sets holds. Examples of

$$ND(\Gamma) \supseteq NE(\Gamma), \ ND(\Gamma) \cap NE(\Gamma) = \emptyset$$

and

$$ND(\Gamma) \nsubseteq NE(\Gamma) \ \wedge \ ND(\Gamma) \nsupseteq NE(\Gamma) \ \wedge \ ND(\Gamma) \cap NE(\Gamma) \neq \emptyset$$

complete the section.

Two Classes of Path Player Games Where $ND(\Gamma)$ Is Contained in $NE(\Gamma)$

For the first class of path player games, we require strictly increasing cost functions. For the second class we need no restriction on the cost function but a path-disjoint network. In both classes we find instances, where $ND(\Gamma)$ is a true subset of $NE(\Gamma)$. We recall Theorem 2.45, which states that in a game with strictly increasing cost functions c_e on all edges $e \in E$ and $\omega_P = 0 \ \forall \ P \in \mathcal{P}$ it holds that

$$f \in NE(\Gamma) \quad \Leftrightarrow \quad \sum_{P \in \mathcal{P}} f_P = r.$$

Theorem 2.70. *Let Γ be a path player game with strictly increasing cost functions c_e for all $e \in E$ and $\omega_P = 0 \ \forall \ P \in \mathcal{P}$. It holds:*

$$ND(\Gamma) \subseteq NE(\Gamma).$$

Proof. Consider a nondominated flow f^*. Assume, f^* is not an equilibrium; that is, $\sum_{P \in \mathcal{P}} f_P^* < r$. Take an arbitrary path $\hat{P}$ and set

$$f_{\hat{P}} = f_{\hat{P}}^* + r - \sum_{P \in \mathcal{P}} f_P^* > f_{\hat{P}}^*.$$

Due to the strictly increasing cost functions it follows that

$$b_{\hat{P}}(f) > b_{\hat{P}}(f^*) \quad \text{and} \quad b_P(f) \geq b_P(f^*) \ \forall \ P \in \mathcal{P} \setminus \{\hat{P}\}.$$

Thus, $b_P(f) \gneq b_P(f^*)$ which contradicts $f^* \in ND(\Gamma)$. Hence, $f^* \in NE(\Gamma)$. $\qquad\square$

The following example illustrates that the reverse of Theorem 2.70 does not hold in general.

Example 2.71. Consider the game illustrated in Figure 2.27. There are two paths going from s to t. Path P_1 shares all its edges with P_2, whereas P_2 owns some edges exclusively. Consider a flow f^* being an equilibrium; that is, $\sum_{P \in \mathcal{P}} f_P = r$. Set $f_{P_1} = f_{P_1}^* - \delta$ and $f_{P_2} = f_{P_2}^* + \delta$. Due to the strictly increasing cost functions, the benefit of P_2 will increase, whereas the benefit of P_1 is unchanged; that is, f dominates f^*: $b(f) \gneq b(f^*)$.

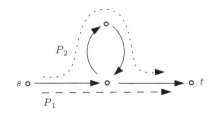

Fig. 2.27. Game network Example 2.71.

Using the example and Theorem 2.70, we conclude that there exist path player games Γ with strictly increasing cost functions c_e for all $e \in E$ and no security limit, such that $ND(\Gamma) \subsetneq NE(\Gamma)$. Nevertheless, there are games with strictly increasing costs, where the reverse of Theorem 2.70 is true (i.e., where $ND(\Gamma) = NE(\Gamma)$ holds). We discuss these games in the following section, regarding a class of games with $ND(\Gamma) = NE(\Gamma)$.

In the second case we assume a path-disjoint network, where each edge belongs to exactly one path. In such networks it holds that the cost of a path depends only on the flow on that path, such that $c_P(f) = c_P(f_P)$.

Lemma 2.72. *Consider a game on a path-disjoint network with continuous cost functions c_e. For each feasible flow f that is no equilibrium there is a feasible equilibrium f^* dominating f.*

Proof. Consider a feasible flow f that is no equilibrium; that is, $\exists\, P : f_P \notin f_P^{\max}$. We construct the required equilibrium f^* by repeating the following iteration until $f_P \in f_P^{\max} \ \forall\, P \in \mathcal{P}$. Note that $f_{\bar{P}}^{\max} \neq \emptyset$ for continuous cost functions (see Lemma 2.24).

Iteration: Choose some $\bar{P} : f_{\bar{P}} \notin f_{\bar{P}}^{\max}$. Set f' such that $f'_P = f_P \ \forall\, P \neq \hat{P}$ and $f'_{\bar{P}} \in f_{\bar{P}}^{\max}$. Set $f := f'$ and repeat.

Due to the path-disjoint network $c_P(f) = c_P(f_P)$ holds for all $P \in \mathcal{P}$. By choosing $f_{\bar{P}} \in f_{\bar{P}}^{\max}$, it is not possible that a player will create an infeasible flow in any iteration. Thus, in each iteration and for all P it holds that $d_P \geq f_P$. It follows that the benefit $b_P(f)$ will not change in any iteration where P is not chosen, although f_{-P} might change. The same is true for all flows smaller than f_P and hence,

$$b_P(f'_{-P}, x) = b_P(f_{-P}, x) \ \forall\, x \leq f_P \wedge\ P \neq \bar{P}. \tag{2.19}$$

Thus, we have $b_P(f') = b_P(f)$ for all $P \neq \bar{P}$. Because $b_{\bar{P}}(f') > b_{\hat{P}}(f)$, it follows that f' dominates f and, consequently, dominates the flows of all previous iterations.

In each iteration either $f'_{\bar{P}} < f_{\bar{P}}$ or $f'_{\bar{P}} > f_{\bar{P}}$ holds.

Claim 1: $f'_{\bar{P}} < f_{\bar{P}}$ can only occur when $\bar{P}$ is chosen for the first time.

Proof of Claim 1: Consider an iteration where $\bar{P}$ is chosen the second time or later, $f_{\bar{P}}$ is the flow before this iteration. Suppose $\exists \tilde{f}_{\bar{P}} < f_{\bar{P}}$ such that $b_{\hat{P}}(\tilde{f}_{\bar{P}}) > b_{\hat{P}}(f_{\bar{P}})$. Due to (2.19) this is a contradiction to $f_{\bar{P}} \in f_{\bar{P}}^{\max}$ in the previous iteration where $\bar{P}$ had been chosen; that is, $f'_{\bar{P}} \geq f_{\bar{P}}$.

Claim 2: If after $|\mathcal{P}|$ iterations no f_P has decreased, the process has found an equilibrium.

Proof of Claim 2: For any path P that has been chosen, the following holds during the $|\mathcal{P}|$ iterations. For all $P_k \neq P$ the flow f_{P_k} has not decreased, thus d_P has not increased. Due to (2.19) and as $f_P^{\max}$ is taken over a feasible region that is smaller than or equal to the iteration where P had been chosen, f_P will stay in $f_P^{\max}$ and will be not chosen again. The process will stop after at most $|\mathcal{P}|$ iterations with $f_P \in f_P^{\max} \ \forall\, P \in \mathcal{P}$.

From Claim 1 and Claim 2 we can follow that after at most $2|\mathcal{P}|$ iterations the iteration stops with an equilibrium. Because this equilibrium dominates each previous flow it also dominates the starting flow f. □

The next result follows immediately.

Theorem 2.73. *Let Γ be a path player game on a path-disjoint network with continuous cost functions. Then $ND(\Gamma) \subseteq NE(\Gamma)$.*

Example 2.69 shows that the reverse of Theorem 2.73 is not true. There a path-disjoint network has been observed and a feasible equilibrium f^* has been found that   is dominated such that $ND(\Gamma) \neq NE(\Gamma)$ may hold for instances of games on path-disjoint networks. Thus, as in games with strictly increasing cost functions, in games on path-disjoint networks there are also instances where $ND(\Gamma) \subsetneqq NE(\Gamma)$. However, there are also instances where $ND(\Gamma) = NE(\Gamma)$; see the following paragraph.

A Class of Path Player Games Where $ND(\Gamma)$ Equals $NE(\Gamma)$

It is a nice property of a game if the set of equilibria equals the set of nondominated flows. Then we can be sure that each equilibrium is "acceptable" in the sense that we can find no dominating flow. On the other hand, each nondominated situation is stable, as it is an equilibrium. Path player games with $ND(\Gamma) = NE(\Gamma)$ are found in the class of games on path-disjoint networks with strictly increasing cost functions and no security limit.

Lemma 2.74. *Let Γ be a path player game on a path-disjoint network with strictly increasing cost functions and $\omega_P = 0 \; \forall P \in \mathcal{P}$. Then $ND(\Gamma) = NE(\Gamma)$.*

Proof. $ND(\Gamma) \subseteq NE(\Gamma)$ is true by Theorem 2.70.
For the reverse conclusion, consider a feasible equilibrium f^*. According to Theorem 2.45 $\sum_{P \in \mathcal{P}} f_P^* = r$. Assume f^* is dominated; that is, $\exists f : b(f) \gneqq b(f^*)$

$$\Rightarrow \exists \, P : b_P(f) > b_P(f^*)$$
$$\Rightarrow f_P > f_P^* \qquad \text{[as } c_e(f_e) \text{ strictly increasing and paths are disjoint]}$$
$$\Rightarrow \exists \, \hat{P} : f_{\hat{P}} < f_{\hat{P}}^* \qquad \text{[as otherwise feasibility would be violated]}$$
$$\Rightarrow b_{\hat{P}}(f) < b_{\hat{P}}(f^*)$$

which contradicts the assumption. Hence:

$$\nexists f : b(f) \gneqq b(f^*) \quad \Rightarrow \quad f^* \in ND(\Gamma). \qquad \square$$

Further Relations Between the Sets $ND(\Gamma)$ and $NE(\Gamma)$

The following relations between the sets $ND(\Gamma)$ and $NE(\Gamma)$ exist and are demonstrated by examples.

Case 1: $ND(\Gamma) \nsubseteq NE(\Gamma) \; \wedge \; ND(\Gamma) \nsupseteq NE(\Gamma) \; \wedge \; ND(\Gamma) \cap NE(\Gamma) \neq \emptyset$

For general games Γ it is possible that neither the nondominated flows nor the equilibria are subsets of each other, but intersection points are existing. See the following example.

Example 2.75. Consider the game described by the graph in Figure 2.28 and the cost functions c_{e_1} and c_{e_2} illustrated in Figures 2.29 and 2.30. The cost functions on the remaining edges are constant zero. In this game there are three paths. The benefit of P_1 depends on the sum of all flows and is illustrated in Figure 2.31. The benefits of paths P_2 and P_3 are equal and depend on the sum $f_{P_2} + f_{P_3}$ and on f_{P_1}. Therefore b_{P_1} and b_{P_2} are two-dimensional functions; we illustrate them for $f_{P_1} = 0$ in Figure 2.32.

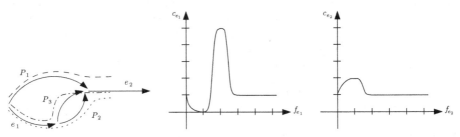

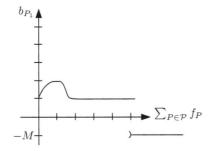

Fig. 2.28. Game graph.   **Fig. 2.29.** Cost of e_1.   **Fig. 2.30.** Cost of e_2.

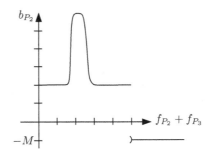

Fig. 2.31. Cost of P_1.   **Fig. 2.32.** Cost of P_2, P_3 for $f_{P_1} = 0$.

Consider the three flows: $f^* = (0, 2.5, 2.5)$, $f^{**} = (0, 1, 1)$, and $f = (1, 0, 0)$ with the benefits $b(f^*) = (1, 2, 2)$, $b(f^{**}) = (1, 6, 6)$, and $b(f) = (2, 3, 3)$.

The flows f^* and f^{**} are equilibria; f is not. In addition f^* is dominated by f and by f^{**}. Thus, f^* is in $NE(\Gamma)$ but not in $ND(\Gamma)$.

The flow f is not dominated by f^{**} and nor by any other flow. A flow $\bar{f}$ that would dominate f needs to have $b(\bar{f}) \gneqq (2, 3, 3)$. As the benefit of player P_1 will never exceed 2, $b_{P_1}(\bar{f}) = 2$ has to hold. The benefit $b_{P_1}(\bar{f}) = 2$ is only obtained if $\sum_{P \in \mathcal{P}} \bar{f}_P = 1$.

$$\Rightarrow \bar{f}_{P_2}, \bar{f}_{P_3} \leq 1$$
$$\Rightarrow \forall\, i = 2, 3 : b_{P_i}(\bar{f}) = c_{P_i}(\bar{f}) = c_{e_1}(f_{e_1}) + 2 \leq 1 + 2$$
$$\Rightarrow \nexists\, \bar{f} : b(\bar{f}) \gneqq (2, 3, 3),$$

and thus, f is undominated; that is, $f \in ND(\Gamma) \,\wedge\, f \notin NE(\Gamma)$.

Case 2: $ND(\Gamma) \supset NE(\Gamma)$

There are path player games where the set of equilibria is contained in the set of nondominated flows as the following example illustrates.

Example 2.76. Consider the game illustrated in Figure 2.33. Each of the two paths connecting source and sink consists of two edges. The exclusive edge of player 1 has cost $c_{e_1}(f_1) = -3f_1$ and that of player two has cost $c_{e_2}(f_2) = -10f_2$. The common used edge has cost $c_{e_3}(f) = 2(f_1 + f_2)$.

As in each game situation, both players will have incentive to decrease their own flows, the unique equilibrium of that game is $f^* = (0, 0)$; see Figure 2.34. It can be verified that the set of nondominated flows is given by

$$ND(\Gamma) = \{f : f_1 = 0 \,\wedge\, 0 \leq f_2 \leq r\} \,\cup\, \{f : 0 \leq f_1 \leq r \,\wedge\, f_2 = 0\}.$$

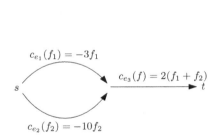

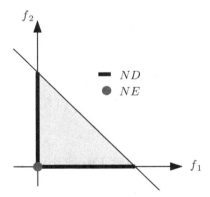

Fig. 2.33. Example 2.76.

Fig. 2.34. $ND(\Gamma) \supset NE(\Gamma)$.

Case 3: $ND(\Gamma) \cap NE(\Gamma) = \emptyset$

A game where each equilibrium is dominated is disadvantageous for the players. A game of that nature is, for example, the Prisoner's Dilemma. Also in path player games we can make out such situations, as the following example illustrates.

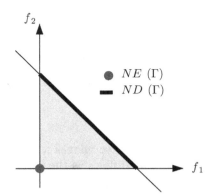

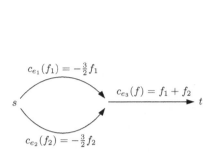

Fig. 2.35. Example 2.77.   **Fig. 2.36.** $ND(\Gamma) \cap NE(\Gamma) = \emptyset$.

Example 2.77. Consider the game illustrated by the graph in Figure 2.35. We have two paths 1 and 2 from source to sink; each path possesses an exclusive and a common used edge. The costs of the exclusive edges are $c_{e_1}(f_1) = -3/2f_1$ and $c_{e_2}(f_2) = -3/2f_2$. The common used edge has cost $c_{e_3}(f_1, f_2) = f_1 + f_2$. Thus we have costs on the paths 1 and 2 of $c_1 = f_2 - 1/2f_1$ and $c_2 = f_1 - 1/2f_2$. We have no security limit; that is, $\omega_P = 0$ for $P \in \{1, 2\}$. Furthermore, the flow rate is $r = 1$.

In this game, the unique feasible equilibrium is $f^* = (0, 0)$; see Figure 2.36. Assume there would be another feasible equilibrium where player 1 (without loss of generality) sends a positive flow $f_1 > 0$. Then $b_1(f_1 - \varepsilon, f_2) = f_2 - 1/2(f_1 - \varepsilon) < f_2 - 1/2f_1 = b_1(f_1, f_2)$ for any fixed $f_2 \geq 0$. It follows that (f_1, f_2) is no equilibrium. The set of nondominated flows is given by all flows that maximize $f_1 + f_2$; that is,

$$ND = \{f = (f_1, f_2) : f_1 + f_2 = r, f_1 \geq 0, f_2 \geq 0\}.$$

2.4 Potential Functions for Path Player Games

2.4.1 Introduction

In this section we investigate *potential functions* for path player games. For a noncooperative game, a potential function is a single (but not necessarily unique) function that represents for each player the change of payoff due to the change of her strategy. In addition, a potential function does not depend on the player. Existence of potential functions is given only for a few classes of games. However, potential functions turn out to be interesting for proving the existence of equilibria in a game. Furthermore, algorithms for the computation of equilibria can be designed based on the existence of potential functions.

We start the investigation of potential functions by considering only feasible flows. This is motivated by the fact that if we consider the complete set of flows $f \in \mathbb{R}_+^n$, the situation gets more complicated due to the noncontinuity of the benefit function. Thus, in terms of introducing this theory, it is more convenient to start with the investigation of feasible flows and to extend this concept later on to the complete set of flows. By considering only feasible flows, we obtain a different type of game, namely a *generalized path player game*, which is characterized by strategy sets that are restricted by the chosen strategies of the competitors. We introduce this concept in more detail in Section 2.4.2. As the classical definition of a potential function is not applicable to restricted strategy sets, we propose a new type of potential function, called the *restricted potential function*. We show that generalized path player games possess exact restricted potential functions. Furthermore, we develop an algorithmic approach to compute feasible equilibria by solving an optimization problem. In Section 2.4.3, we consider the complete strategy set, including the infeasible flows. Due to the structure of the benefit function, an exact potential does not exist here. Nevertheless, we prove the existence of ordinal potential functions, a weaker form of potentials. In a third approach in Section 2.4.4 we extend the benefit function such that even an exact potential function for the complete strategy set can be found. Finally, in Section 2.4.5, computation of equilibria by using a greedy approach is discussed. We present classes of path player games, where equilibria can be found within a finite sequence of greedy improvement steps.

Potential functions were first mentioned by Rosenthal [Ros73], who used a potential function to prove the existence of pure-strategy equilibria in finite congestion games. Monderer and Shapley introduced in [MS96] several definitions of potentials and characterized games with potential functions by sequences of strategy vectors. Facchini et al. [FMBT97] presented a characterization of potential games by a sum of a coordination and a dummy game. Furthermore, they extended the concept of congestion games to incomplete information and yielded Bayesian potential games. Voorneveld and Norde [VN97] presented a characterization of ordinal potential games by using weak improvement cycles (cycles where an improvement takes place in at least one step) which do not exist in these games.

As path player games are infinite, our special interest belongs to *infinite potential games*. Monderer and Shapley presented in [MS96] sufficient conditions for both the existence of approximate equilibria and the existence of equilibria in infinite potential games. The former is satisfied by path player games, as the benefit functions are bounded; the latter is not, because we have noncontinuous benefits. Kukushkin [Kuk99] proved the existence of equilibria in infinite ordinal potential games with compact strategy sets. Norde and Tijs [NT98] considered games where one or more players have infinite strategy sets and present sufficient conditions for weak determinedness of these games. A game is weakly determined if it provides (ε, k)-equilibria; that is, equilibria where each player is "reasonably" satisfied. That means he either cannot improve his benefit more than an ε or is already gaining a benefit higher than k. In [Voo97], Voorneveld extended results by Norde and Tijs, which were given for exact potential games to generalized ordinal potential games.

Some research has been done linking the field of potential games with other fields in game theory. Potential functions in cooperative games were considered, for example, by Bilbao [Bil98] and Driessen and Radzika [DR02]. Ui and Slikker et al. in [Ui00, SDNT00, Sli01] used the relation of potential games and the Shapley value of a cooperative game. Games with incomplete information and robustness of equilibria in potential games were studied by Ui and Morris in [Ui01, MU05]. Evolutionary processes were considered by Sandholm [San01] where infinite player sets were given. In Baron et al. [BDHS02] evolutionary dynamics were considered under stochastic perturbations. Mallozzi et al. [MTV00] combined the concept of hierarchical games, such as Stackelberg games, with potential functions and introduced a hierarchical potential game. Kukushkin [Kuk02] presented potential functions for games with perfect information and the relation to subgame perfect equilibrium. In [PPT07] the authors introduced a multicriteria version of potential games and extended results from singlecriteria case to this field.

Variations of the classical definition of potential games presented in [MS96] have been studied, for example, in the following publications. In [Voo00] Voorneveld proposed *best-response potential games*, a class of games containing the class of ordinal potential games. Best-response potential functions provide a sufficient condition for the existence of equilibria. A recent approach was presented in Monderer [Mon07] where multiple player types were considered in *q-potential games*.

Finally, we mention that potential functions were also studied for cooperative games (see, e.g., [HMC89, Bil98]). As we are only considering noncooperative games, this theory is not considered further here.

Definition 2.78. *Let* $w = (w_P)_{P \in \mathcal{P}}$ *be a vector of weights with* $w_P > 0 \; \forall \; P \in \mathcal{P}$. *A function* $\Pi : \mathbb{R}_+^{|P|} \to \mathbb{R}$ *is a* w-potential *for a game* Γ *if for every* $P \in \mathcal{P}$, *for every* $f_{-P} \in \mathbb{R}_+^{|\mathcal{P}|-1}$, *and for every* $x, z \in \mathbb{R}_+$ *it holds:*

$$b_P(f_{-P}, x) - b_P(f_{-P}, z) = w_P \left(\Pi(f_{-P}, x) - \Pi(f_{-P}, z) \right).$$

Γ *is called a* w-potential game *if it admits a* w-potential.

For $w = \mathbf{1}_{|\mathcal{P}|}$ we receive the strongest form of a potential, an *exact potential*. A weaker form of a potential function is the *ordinal potential*, introduced next.

Definition 2.79. *A function* $\Pi : \mathbb{R}_+^{|P|} \to \mathbb{R}$ *is called an* ordinal potential *for* Γ *if for every* $P \in \mathcal{P}$, *for every* $f_{-P} \in \mathbb{R}_+^{|\mathcal{P}|-1}$, *and for every* $x, z \in \mathbb{R}_+$,

$$b_P(f_{-P}, x) - b_P(f_{-P}, z) > 0 \quad \Leftrightarrow \quad \Pi(f_{-P}, x) - \Pi(f_{-P}, z) > 0.$$

Γ *is called an* ordinal potential game *if it admits an ordinal potential.*

The definition of an ordinal potential can be modified to obtain a *generalized ordinal potential*.

Definition 2.80. *A function* $\Pi : \mathbb{R}_+^{|P|} \to \mathbb{R}$ *is called a* generalized ordinal potential *for* Γ *if for every* $P \in \mathcal{P}$, *for every* $f_{-P} \in \mathbb{R}_+^{|\mathcal{P}|-1}$, *and for every* x, $z \in \mathbb{R}_+$

$$b_P(f_{-P}, x) - b_P(f_{-P}, z) > 0 \quad \Rightarrow \quad \Pi(f_{-P}, x) - \Pi(f_{-P}, z) > 0.$$

Γ *is called a* generalized ordinal potential game *if it admits a generalized ordinal potential.*

It is clear by the definition of the types of potential functions that an exact potential Π in game Γ induces a w-potential in Γ. A w-potential itself then induces an ordinal potential in Γ, which finally induces a generalized potential in Γ. The following lemma is true by the definition of an equilibrium (see Definition 2.26).

Lemma 2.81 ([MS96]). *Consider an ordinal potential game* Γ. *The flow* f^* *is an equilibrium in* Γ *if and only if for every* $P \in \mathcal{P}$ *and for every* f_P *it holds:*

$$\Pi(f_{-P}^*, f_P^*) \geq \Pi(f_{-P}^*, f_P).$$

It follows that if we can find a maximizer for the ordinal potential function $\Pi(f)$, then the equilibrium for Γ exists in pure strategies. Monderer and Shapley [MS96] use that fact to draw the conclusion that each finite ordinal potential game has a pure-strategy equilibrium.

Thus, knowing an (ordinal) potential helps to identify equilibria in a game. It is therefore interesting to have ways to check if a game is a potential game and to determine the potentials themselves. In the following material we investigate *strategy sequences* for this purpose.

Definition 2.82. *A* strategy sequence $\varphi = (f^0, f^1, \ldots, f^k, \ldots)$, *we say simply* sequence, *in a game is given as an ordered sequence of flows* f^k *that satisfies the following. For every* $k \geq 1$ *there is a unique player* $P(k)$ *such that for*

$$f^{k-1} = \left(f_{-P(k)}^{k-1}, f_{P(k)}^{k-1} \right) \quad and \quad f^k = \left(f_{-P(k)}^k, f_{P(k)}^k \right)$$

we have

$$f_{-P(k)}^{k-1} = f_{-P(k)}^k \quad and \quad f_{P(k)}^{k-1} \neq f_{P(k)}^k.$$

We call $P(k)$ *the* active player *and the movement from* f^{k-1} *to* f^k *the* k*th* step *in* φ.

A sequence is called an improvement sequence *if for every* $k \geq 1$ *it holds that*

$$b_{P(k)}\left(f^k \right) > b_{P(k)}\left(f^{k-1} \right). \tag{2.20}$$

A game Γ *satisfies the* finite improvement property (FIP) *if every improvement sequence is finite.*

For a sequence φ, we call f^0 its initial flow *and, if φ is finite, f^N its* terminal flow. *Furthermore, we say that a finite φ* connects f^0 *and f^N. The* length *of a finite sequence $\varphi = (f^0, f^1, \ldots, f^N)$ is given by $l(\varphi) = N$.*

Note, that Monderer and Shapley called the "strategy sequence" simply "path". Because this term in our work is already occupied by the paths that the players own in the network, we use the term "sequence".

A finite improvement sequence that ends because no improvement step is possible any more (a so-called *maximal sequence*) provides an equilibrium as the terminal flow. If a game satisfies FIP we can hence use improvement sequences to determine equilibria.

Lemma 2.83 ([MS96]). *Every finite ordinal potential game satisfies FIP.*

To prove this, note that by (2.20) for each improvement sequence, the potential function values of the flows have to increase strictly in each step. As the set of strategies is finite, each improvement sequence has to be finite.

Unfortunately, although we show in the following sections that path player games have potential functions, we cannot apply that result to them. Path player games have an infinite number of strategies, and thus the improvement of benefit in a step may become arbitrarily small, which may lead to infinite improvement sequences. An example of such an infinite improvement sequence is given in Example 2.111 on page 79. See Section 2.4.5 for the analysis of other approaches to create sequences that yield equilibria or approximate equilibria.

Definition 2.84. *The* cost *of a finite sequence $\varphi = \left(f^0, f^1, \ldots, f^N\right)$ is given by*

$$I(\varphi) = \sum_{k=1}^{N} \left[b_{P(k)}\left(f^k\right) - b_{P(k)}\left(f^{k-1}\right) \right]. \tag{2.21}$$

Corollary 2.85. *Let $\varphi^{-1} = \left(f^N, f^{N-1}, \ldots, f^0\right)$ be the sequence with reverse ordering of φ. Then,*

$$I(\varphi) = -I(\varphi^{-1}).$$

Definition 2.86. *A sequence is* closed *if $f^0 = f^N$ and a closed sequence is* simple *if $f^\ell \neq f^k$ for all $\ell \neq k$ and $0 \leq \ell, k \leq N - 1$.*

Lemma 2.87 ([MS96]). *Consider a game Γ. The following statements are equivalent.*

Γ *is an exact potential game.* $\hfill (2.22)$

$I(\varphi) = 0$ *for every finite closed sequence φ.* $\hfill (2.23)$

$I(\varphi) = 0$ *for every finite simple closed sequence φ.* $\hfill (2.24)$

$I(\varphi) = 0$ *for every simple closed sequence φ of length four.* $\hfill (2.25)$

An exact potential of Γ is given by fixing a flow $\bar{f} \in \mathbb{F}$ and defining $\Pi(f) = I(\varphi) \; \forall f \in \mathbb{F}$, where φ is a sequence connecting $\bar{f}$ and f. Note that $\Pi(f)$ is well defined; that is, $I(\varphi_1) = I(\varphi_2)$ if φ_1 and φ_2 have the same initial and terminal flow. In [MS96] it is in addition shown for exact potential games Γ that taking any two exact potentials Π^1 and Π^2 there is an constant c such that $\Pi^1(f) - \Pi^2(f) = c$ for all flows f.

For further consideration, we assume in this section to have path player games without a security limit; that is, $\omega_P = 0 \; \forall \; P \in \mathcal{P}$. Later on, we develop potential functions for games on polyhedra, a generalization of path player games, considered in Chapter 3. There, we propose ways also to include a general security limit.

In terms of path player games and potential functions, we start with a negative result. Path player games are not exact potential games in general, which is shown in the following simple example.

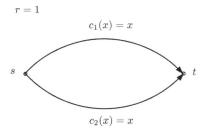

$r = 1$

$c_1(x) = x$

s

t

$c_2(x) = x$

Fig. 2.37. Example 2.88.

Example 2.88. Consider the path player game illustrated in Figure 2.37 and the sequences

$$\varphi_1 = ((0,0), (0, 0.75), (0.5, 0.75)) \quad \text{and} \quad \varphi_2 = ((0,0), (0.5, 0), (0.5, 0.75)) \, .$$

We have: $I(\varphi_1) = (0.75 - 0) + (-M - 0) = 0.75 - M$ and $I(\varphi_2) = (0.5 - 0) + (-M - 0) = 0.5 - M$ and hence $I(\varphi_1) \neq I(\varphi_2)$. The closed sequence which is obtained by connecting φ_1 and φ_2^{-1} has cost $0.75 - 0.5 - M + M = 0.25$. By Lemma 2.87, the presented game is not an exact potential game.

Nevertheless, we are able to develop other approaches for determining potential functions in path player games. In Section 2.4.2, we define an exact restricted potential function that exists in the generalized version of the path player game. For the original path player game, we propose an ordinal potential in Section 2.4.3. Furthermore, an extension of the original benefit function allows an exact potential for the path player game, which we analyze in Section 2.4.4.

2.4.2 Exact Restricted Potential for the Generalized Path Player Game

In this section, we introduce a new definition of potential functions, valid for games with restricted strategy sets. With this new notation, we are able to obtain an exact (restricted) potential function for the path player game. For this approach we consider only the set of feasible flows. We allow a player to choose f_P only from $[0, d_P(f_{-P})]$. By feasibility, the benefit is then given by $b_P(f) = c_P(f)$. Games where the strategy sets of the players are dependent on the strategies of the competitors are called *generalized Nash equilibrium (GNE) games*, but are also known under several other names. See Chapter 3 for a more detailed introduction into this field. Consequently, we call a path player game, where the strategy set of each player is restricted to $[0, d_P]$, a *generalized path player game*.

Definition 2.89. *In a generalized path player game, a feasible flow f^* is a generalized equilibrium if and only if for all paths $P \in \mathcal{P}$ and for all $f_P \in [0, d_P(f^*_{-P})]$, it holds that*

$$b_P(f^*_{-P}, f^*_P) \geq b_P(f^*_{-P}, f_P).$$

The investigation of generalized path player games is not only interesting for studying potential functions. It also makes sense if we are interested in feasible equilibria, as the set of feasible equilibria in a path player game and the set of generalized equilibria in the corresponding generalized path player game are equal.

Theorem 2.90. *Consider a path player game Γ and the corresponding generalized path player game $\hat{\Gamma}$. A flow f^* is a feasible equilibrium in Γ if and only if f^* is a generalized equilibrium in $\hat{\Gamma}$.*

Proof.
Part (a) f^* is a feasible equilibrium in Γ $\Rightarrow$ f^* is a generalized equilibrium in $\hat{\Gamma}$.
As f^* is a feasible equilibrium in Γ, we have

$$\forall\, P \in \mathcal{P},\ \forall\, f_P \geq 0: \qquad b_P(f^*_{-P}, f^*_P) \geq b_P(f^*_{-P}, f_P)$$
$$\Rightarrow \forall\, P \in \mathcal{P},\ \forall\, f_P \in [0, d_P(f^*_{-P})]: b_P(f^*_{-P}, f^*_P) \geq b_P(f^*_{-P}, f_P).$$

As f^* is feasible, it is a generalized equilibrium in $\hat{\Gamma}$ by definition.
Part (b) f^* is a generalized equilibrium in $\hat{\Gamma}$ $\Rightarrow$ f^* is a feasible equilibrium in Γ.
As f^* is a generalized equilibrium in $\hat{\Gamma}$, we have $\forall\, P \in \mathcal{P},\ \forall\, f_P \in [0, d_P(f^*_{-P})]$:

$$b_P(f^*_{-P}, f^*_P) \geq b_P(f^*_{-P}, f_P) > -M = b_P(f^*_{-P}, \tilde{f}_P),$$

for $\tilde{f}_P > d_P(f^*_{-P})$. Thus,

$$\forall\, P \in \mathcal{P},\ \forall\, f_P \geq 0 : b_P(f^*_{-P}, f^*_P) \geq b_P(f^*_{-P}, f_P).$$

As f^* is feasible, the claim follows. □

Because we are restricting ourselves to feasible flows, the definition of potentials needs an adjustment. In the definition of a classical potential it is not considered that strategy sets may be restricted.

Definition 2.91. Let $w = (w_P)_{P \in \mathcal{P}}$ be the vector of weights with $w_P > 0\ \forall\, P \in \mathcal{P}$. A function $\Pi : \mathbb{F} \to \mathbb{R}$ is a restricted w-potential for a generalized path player game Γ if for every $f \in \mathbb{F}$, for every $P \in \mathcal{P}$, and for all $x,\, z \in [0, d_P(f_{-P}]$ it holds:

$$b_P(f_{-P}, x) - b_P(f_{-P}, z) = w_P\left(\Pi(f_{-P}, x) - \Pi(f_{-P}, z)\right). \tag{2.26}$$

Consequently, a path player game Γ is called a restricted w-potential game *if it admits a restricted w-potential. For $w = \mathbf{1}_{|\mathcal{P}|}$ we obtain an* exact *restricted potential function.*

This definition is extended to *games on polyhedra*, which is an instance of generalized equilibrium games; see Definition 3.38 on page 111. The following statement has already been observed for the classical definition of potential games in Lemma 2.81, page 61, and is now given for generalized path player games.[5]

Lemma 2.92. *Consider a generalized path player game Γ which is a restricted w-potential game. The flow f^* is a feasible generalized equilibrium in Γ if and only if for every $P \in \mathcal{P}$ and for every $f_P \in [0, d_P(f^*_{-P})]$ it holds:*

$$\Pi(f^*_{-P}, f^*_P) \geq \Pi(f^*_{-P}, f_P).$$

Definition 2.93. *A sequence $\varphi = (f^0, \ldots, f^N)$ is called* feasible *if f^k is feasible for every $0 \leq k \leq N$.*

To prove the existence of exact restricted potential functions for generalized path player games (Theorem 2.96), which is the main result of this section, we need the following theorem.

Theorem 2.94. *The following statements are equivalent.*

Γ *is an exact restricted potential game.* (2.27)

$I(\varphi) = 0$ *for every finite closed feasible sequence φ.* (2.28)

$I(\varphi) = 0$ *for every finite simple closed feasible sequence φ.* (2.29)

$I(\varphi) = 0$ *for every simple closed feasible sequence φ of length 4.* (2.30)

---

[5] The statement of Lemma 2.92 would still be true if we defined an "ordinal restricted potential game" along the lines of Definition 2.79. We omit this definition, as it is not relevant for the following investigations.

An exact restricted potential of Γ is given by fixing a feasible flow $\bar{f}$ and defining $\Pi(f) = I(\varphi)$ for all $f \in \mathbb{F}$ where φ is a feasible sequence connecting $\bar{f}$ and f. Note that $\Pi(f)$ is well defined, as $I(\varphi_1) = I(\varphi_2)$ holds for φ_1 and φ_2 having the same initial and terminal flow.

Proof. To prove the thesis of this lemma we follow an argument similar to the one in [MS96] for the proof of Lemma 2.87, but applied to sets of strategies that are mutually dependent (the strategies of the different players are linked by constraints) and to our definition of restricted potential functions. The main novelty of our proof is given in Part (c) where the feasibility of the considered improvement sequence has to be ensured.

First, note that (2.28) implies (2.29) which implies (2.30). It remains to show that (2.27) is equivalent to (2.28) and (2.30) implies (2.28).

Part (a) $(2.27) \Rightarrow (2.28)$.

Let Π be an exact restricted potential for Γ. Consider a feasible closed sequence φ. By the definition of an exact restricted potential, the cost of a sequence is given by

$$I(\varphi) = \sum_{k=1}^{N} \left[\Pi\left(f^k\right) - \Pi\left(f^{k-1}\right) \right] = \Pi\left(f^N\right) - \Pi\left(f^0\right) = 0.$$

Part (b) $(2.28) \Rightarrow (2.27)$.

We assume that $I(\varphi) = 0$ holds for any finite closed feasible sequence φ. Now consider two feasible flows $\bar{f}, f \in \mathbb{F}$. Any finite feasible sequence that has $\bar{f}$ as initial and terminal flow and contains f has cost 0 by assumption. Thus, using Corollary 2.85 it can be seen that all feasible sequences connecting $\bar{f}$ and f have to have equal cost.

We define: for $f \in \mathbb{F}$, we set $\Pi(f) = I(\varphi)$, for all feasible φ connecting a fixed feasible $\bar{f}$ with f. It remains to show that $\Pi(f)$ is an exact restricted potential in Γ. Consider any $f \in \mathbb{F}$ and any $P \in \mathcal{P}$. Furthermore, consider $x, z \in [0, d_P(f_{-P})]$. Let $\varphi_1 = (\bar{f}, f^1, \ldots, (f_{-P}, x))$ be a feasible sequence connecting $\bar{f}$ and (f_{-P}, x). Set $\varphi_2 = (\bar{f}, f^1, \ldots, (f_{-P}, z))$. It follows that $\Pi(f_{-P}, x) - \Pi(f_{-P}, z) = I(\varphi_1) - I(\varphi_2) = b_P(f_{-P}, x) - b_P(f_{-P}, z)$, from which we conclude that Π is an exact restricted potential.

Part (c) $(2.30) \Rightarrow (2.28)$.

We suppose that $I(\varphi) = 0$ for every simple closed feasible sequence φ of length $l(\varphi) = 4$.

Assume there is a finite closed feasible sequence with nonzero cost and let us consider such a sequence $\varphi = (f^0, \ldots, f^N)$ with minimal length $l(\varphi) > 4$. As the sequence is closed, there is a step q with $f^q_{P(q)} - f^{q-1}_{P(q)} < 0$; that is, player $P(q)$ is decreasing his flow. Without loss of generality, let q be the first step: $q = 1$ and set the active player in the first step $P(q) = P_1$. Because $f^0 = f^N$ there has to be a $2 \le j \le N$ such that $P(j) = P_1$ with $f^q_{P(q)} - f^{q-1}_{P(q)} > 0$; that is, player P_1 has to be active a second time where he increases his flow. For $j = 2$ we obtain a contradiction to the minimality of $l(\varphi)$, due to $I(f^0, f^2, \ldots, f^N) = I(\varphi)$. A similar contradiction can be obtained for $j = N$. Hence, we assume $2 \leq j \leq N - 1$. Consider the flow

$$\hat{f}_j = \left(f^{j-1}_{-\{P_1, P(j+1)\}}, f^{j-1}_{P_1}, f^{j+1}_{P(j+1)} \right)$$

which is obtained by preceding step $j+1$ before step j. Note that $P(j+1) \neq P_1$ holds as otherwise we would have a contradiction to minimality of $l(\varphi)$. Because

$$f^{j-1}_P = f^{j+1}_P \ \forall \ P \notin \{P_1, P(j+1)\}$$

and

$$f^{j-1}_{P_1} < f^j_{P_1} = f^{j+1}_{P_1},$$

we can conclude that $\hat{f}$ is a feasible flow:

$$\sum_{P \in \mathcal{P}} \hat{f}_P = \sum_{P \in \mathcal{P} \backslash \{P_1, P(j+1)\}} f^{j-1}_P + f^{j-1}_{P_1} + f^{j+1}_{P(j+1)}$$

$$< \sum_{P \in \mathcal{P} \backslash \{P_1, P(j+1)\}} f^{j+1}_P + f^{j+1}_{P_1} + f^{j+1}_{P(j+1)}$$

$$= \sum_{P \in \mathcal{P}} f^{j+1}_P$$

$$\leq r,$$

as f^{j+1} is feasible. It can be verified that the simple feasible sequence of length four $(f^{j-1}, f^j, f^{j+1}, \hat{f}^j)$ is closed and has by assumption cost zero. Thus, the feasible sequences (f^{j-1}, f^j, f^{j+1}) and $(f^{j-1}, \hat{f}^j, f^{j+1})$ have equal length, and consequently $l(\varphi) = l(\hat{\varphi})$ with

$$\hat{\varphi} = (f^0, \ldots, f^{j-1}, \hat{f}^j, f^{j+1}, \ldots, f^N)$$

is true. Note that in $\hat{\varphi}$ it holds $P(j + 1) = P_1$. By iteration of this replacement process we obtain a finite closed feasible sequence φ^* with $I(\varphi^*) = I(\varphi) \neq 0$ and $P(N) = P(1) = P_1$, which leads to a contradiction of the minimality assumption of $l(\varphi)$. We conclude that $I(\varphi) = 0$ holds for each finite closed feasible sequence φ. $\qquad\square$

A different representation of the cost functions is needed for the proof of the following theorem. The cost of a path P is given by the sum of the costs on the edges belonging to this path. We distinguish between two types of edges, the ones belonging exclusively to P, and the ones shared with other paths. We recall the definition of exclusively used edges (see Definition 2.42) and add the definition of commonly used edges.

Definition 2.95. *For player P, we define*

$$E_P^{\mathrm{exc}} = \{e : e \in P \ \wedge\ e \notin P_k \ \forall\ P_k \neq P\}$$

to be the set of exclusively used edges *of P, and*

$$E_P^{\mathrm{com}} = \{e : e \in P \ \wedge\ \exists\ P_k \neq P : e \in P_k\}$$

to be the set of common used edges.

As $f_e = f_P$ *holds for exclusively used edges, we obtain the* cost of a path in extensive form*:*

$$c_P(f) = \sum_{e \in P} c_e(f_e) = \sum_{e \in E_P^{\mathrm{exc}}} c_e(f_P) + \sum_{e \in E_P^{\mathrm{com}}} c_e \left(f_P + \sum_{P_k : e \in P_k, P_k \neq P} f_{P_k} \right).$$
$$(2.31)$$

Theorem 2.96. *Generalized path player games are exact restricted potential games.*

Proof. We show that each generalized path player game satisfies property (2.30) of Theorem 2.94. Consider any two active players P_i and P_j that create a simple closed feasible sequence $\varphi = (f^0, f^1, f^2, f^3, f^0)$ of length $l(\varphi) = 4$ by choosing alternating a new strategy and returning to the first strategy afterwards. We denote the set of strategies of the remaining players with $f_{-\{P_i P_j\}}$. The sequence φ is then given by

$$\varphi = ((f_{-\{P_i P_j\}}, f_{P_i}, f_{P_j}), (f_{-\{P_i P_j\}}, \bar{f}_{P_i}, f_{P_j}), (f_{-\{P_i P_j\}}, \bar{f}_{P_i}, \bar{f}_{P_j}),$$
$$(f_{-\{P_i P_j\}}, f_{P_i}, \bar{f}_{P_j}), (f_{-\{P_i P_j\}}, f_{P_i}, f_{P_j})).$$

For instance, such a sequence is presented in Figure 2.38. As φ is feasible, we have $b_P(f^k) = c_P(f^k)$ for all P and for all $0 \leq k \leq 3$. We obtain

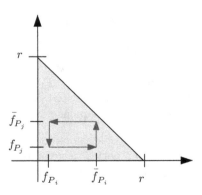

Fig. 2.38. Simple closed feasible sequence of length four.

$$I(\varphi) = \left(c_{P_i}(f^1) - c_{P_i}(f^0)\right) + \left(c_{P_j}(f^2) - c_{P_j}(f^1)\right)$$
$$+ \left(c_{P_i}(f^3) - c_{P_i}(f^2)\right) + \left(c_{P_j}(f^0) - c_{P_j}(f^3)\right).$$

To determine $I(\varphi)$, we need to consider only the cost functions of P_i and P_j, and because $f_{-\{P_iP_j\}}$ is fixed, the influence of $f_{-\{P_iP_j\}}$ in the common edges $E_{P_i}^{\text{com}}$ and $E_{P_j}^{\text{com}}$ can be neglected. Hence, we modify the sets of common and exclusively used edges in the following way,

$$\bar{E}_{P_i}^{\text{exc}} = \{e \in E : e \in P_i \wedge e \notin P_j\} \quad \text{and} \quad \bar{E}_{P_i}^{\text{com}} = \{e \in E : e \in P_i \wedge e \in P_j\}.$$

The values $\bar{E}_{P_j}^{\text{exc}}$ and $\bar{E}_{P_j}^{\text{com}}$ are defined analogously. Because we are only considering the two players P_i and P_j it holds: $\bar{E}_{P_i}^{\text{com}} = \bar{E}_{P_j}^{\text{com}} =: \bar{E}^{\text{com}}$. Thus, to investigate the costs, apart from the exclusive edges of P_i and P_j, we only consider those edges that have P_i and P_j in common, whereas edges that are used in common with the remaining paths carrying invariant flow, are represented in $\bar{E}_{P_i}^{\text{exc}}$ (and $\bar{E}_{P_j}^{\text{exc}}$, resp.). The path cost in extensive form (see (2.31)) is thus rewritten:

$$c_P(f) = \sum_{e \in \bar{E}_P^{\text{exc}}} c_e(f_P) + \sum_{e \in \bar{E}^{\text{com}}} c_e\left(f_P + \sum_{P_k : e \in P_k, P_k \neq P} f_{P_k}\right). \qquad (2.32)$$

We denote $S_{-\{P_iP_j\}}(e) = \sum_{P : e \in P, P \notin \{P_i, P_j\}} f_P$ and obtain:

$$I(\varphi) = \left(\sum_{e \in \bar{E}_{P_i}^{\text{exc}}} c_e(\bar{f}_{P_i}) + \sum_{e \in \bar{E}^{\text{com}}} \left(\bar{f}_{P_i} + f_{P_j} + S_{-\{P_iP_j\}}(e)\right)\right)$$

$$- \left(\sum_{e \in \bar{E}_{P_i}^{\text{exc}}} c_e(f_{P_i}) + \sum_{e \in \bar{E}^{\text{com}}} \left(f_{P_i} + f_{P_j} + S_{-\{P_iP_j\}}(e)\right)\right)$$

$$+ \left(\sum_{e \in \bar{E}_{P_j}^{\text{exc}}} c_e(\bar{f}_{P_j}) + \sum_{e \in \bar{E}^{\text{com}}} \left(\bar{f}_{P_i} + \bar{f}_{P_j} + S_{-\{P_iP_j\}}(e)\right)\right)$$

$$- \left(\sum_{e \in \bar{E}_{P_j}^{\text{exc}}} c_e(f_{P_j}) + \sum_{e \in \bar{E}^{\text{com}}} \left(\bar{f}_{P_i} + f_{P_j} + S_{-\{P_iP_j\}}(e)\right)\right)$$

$$+ \left(\sum_{e \in \bar{E}_{P_i}^{\text{exc}}} c_e(f_{P_i}) + \sum_{e \in \bar{E}^{\text{com}}} \left(f_{P_i} + \bar{f}_{P_j} + S_{-\{P_iP_j\}}(e)\right)\right)$$

$$- \left(\sum_{e \in \bar{E}_{P_i}^{\text{exc}}} c_e(\bar{f}_{P_i}) + \sum_{e \in \bar{E}^{\text{com}}} \left(\bar{f}_{P_i} + \bar{f}_{P_j} + S_{-\{P_iP_j\}}(e)\right)\right)$$

$$+ \left(\sum_{e \in \bar{E}_{P_j}^{\mathrm{exc}}} c_e(f_{P_j}) + \sum_{e \in \bar{E}^{\mathrm{com}}} (f_{P_i} + f_{P_j} + S_{-\{P_i P_j\}}(e)) \right)$$

$$- \left(\sum_{e \in \bar{E}_{P_j}^{\mathrm{exc}}} c_e(\bar{f}_{P_j}) + \sum_{e \in \bar{E}^{\mathrm{com}}} (f_{P_i} + \bar{f}_{P_j} + S_{-\{P_i P_j\}}(e)) \right)$$

$$= \quad 0. \qquad\qquad\qquad\qquad\qquad\qquad\qquad \square$$

Definition 2.97. *Let $n = |\mathcal{P}|$ be the number of players in a generalized path player game. Consider a flow f and a fixed flow $\bar{f}$. A sequence φ of length $l(\varphi) = \nu \leq n+1$ with $\varphi = (\bar{f}, f^1, \dots, f^{\nu-1}, f)$ is called a* direct sequence *from $\bar{f}$ to f, if for all $k = 1, \dots, \nu$ it holds*

$$f_{P(\ell)}^k = f_{P(\ell)} \ \forall \ \ell \leq k \quad and \quad f_{P(\ell)}^k = \bar{f}_{P(\ell)} \ \forall \ \ell > k.$$

In a direct sequence, in each step a unique player $P(k)$ changes its flow from $\bar{f}_{(P_k)}$ to $f_{P(k)}$. After ν steps, f is obtained.

Lemma 2.98. *If f is a feasible flow then each direct sequence connecting the zero-flow $\bar{f} = \mathbf{0}_{|\mathcal{P}|}$ and f is a feasible sequence.*

Proof. As f is feasible it holds $\sum_{P \in \mathcal{P}} f_P \leq r$. For each step $k = 1, \dots, \nu$ it holds

$$\sum_{P \in \mathcal{P}} f_P^k = \sum_{\ell=1}^{k} f_{P(\ell)} + \sum_{\ell=k+1}^{n} 0 \leq \sum_{\ell=1}^{n} f_{P(\ell)} = \sum_{P \in \mathcal{P}} f_P \leq r. \qquad \square$$

Note that Lemma 2.98 does not hold for arbitrary feasible $\bar{f}$, which is illustrated in the next example. Nevertheless, there always exists a feasible sequence connecting two feasible flows $\bar{f}$ and f. We obtain such a sequence by ordering the players such that those players who want to reduce their flow become active first.

Example 2.99. Consider the game represented by Figure 2.37 on page 63. Set $\bar{f} = (0.25, 0.75)$ and $f = (0.5, 0.5)$. The direct sequence $\varphi = ((0.25, 0.75), (0.5, 0.75), (0.5, 0.5))$ is not feasible due to the infeasible flow f^1. A feasible direct sequence is given by $\bar{\varphi} = ((0.25, 0.75), (0.25, 0.5), (0.5, 0.5))$; see Figure 2.39.

By fixing $\bar{f} = \mathbf{0}_{|\mathcal{P}|}$ and using Theorem 2.94 and Lemma 2.98 we derive the following statement.

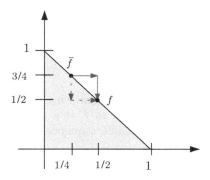

Fig. 2.39. Feasible and infeasible direct sequences.

Lemma 2.100. *Consider a generalized path player game with n players. An exact restricted potential* Π *is given by*

$$
\Pi(f) = I(\varphi) = \left[b_{P(k)} \begin{pmatrix} f_{P(1)} \\ 0 \\ \vdots \\ 0 \end{pmatrix} - b_{P(k)} \begin{pmatrix} 0 \\ \vdots \\ 0 \end{pmatrix} \right]
$$

$$
+ \sum_{k=2}^{n} \left[b_{P(k)} \begin{pmatrix} f_{P(1)} \\ \vdots \\ f_{P(k-1)} \\ f_{P(k)} \\ 0 \\ \vdots \\ 0 \end{pmatrix} - b_{P(k)} \begin{pmatrix} f_{P(1)} \\ \vdots \\ f_{P(k-1)} \\ 0 \\ 0 \\ \vdots \\ 0 \end{pmatrix} \right].
$$

The next theorem provides a shorter representation of the potential function.

Theorem 2.101. *Consider a generalized path player game with n players. An exact restricted potential* Π *is given by*

$$
\Pi(f) = \sum_{e \in E} c_e(f_e).
$$

Before we prove Theorem 2.101, note that it would be an alternative (and much shorter) proof to check that $\Pi(f) = \sum_{e \in E} c_e(f_e)$ satisfies (2.26), the definition of a restricted potential function. This sort of proof is always possible, once you have a potential function given. Nevertheless, the following extensive proof is interesting, as it illustrates the construction of the potential function.

Proof. We show that in an exact restricted potential game, each feasible sequence φ connecting a fixed flow $\bar{f}$, say $\bar{f} = \mathbf{0}_{|\mathcal{P}|}$, and some feasible flow f has cost

$$I(\varphi) = \sum_{e \in E} [c_e(f_e) - c_e(0)].$$

Consider a generalized path player game with n players, a feasible flow f and a direct sequence φ from $\bar{f} = \mathbf{0}_{|\mathcal{P}|}$ to f. It is sufficient to show the claim for direct sequences only, as all feasible sequences connecting $\bar{f}$ and f have equal costs.

As f is feasible, it holds by Lemma 2.98 that φ is feasible too. Denote

$$I_k(\varphi) = b_{P(k)}(f^k) - b_{P(k)}(f^{k-1}) = c_{P(k)}(f^k) - c_{P(k)}(f^{k-1}).$$

Then

$$I_k(\varphi) = \sum_{e \in E_{P(k)}^{\mathrm{exc}}} c_e(f_{P(k)}) + \sum_{e \in E_{P(k)}^{\mathrm{com}}} c_e \left[f_{P(k)} + \sum_{P:e \in P \wedge P \neq P(k)} f_P^k \right]$$
$$- \left(\sum_{e \in E_{P(k)}^{\mathrm{exc}}} c_e(0) + \sum_{e \in E_{P(k)}^{\mathrm{com}}} c_e \left[\sum_{P:e \in P \wedge P \neq P(k)} f_P^k \right] \right).$$

And thus

$$I(\varphi) = \sum_{k=1}^{n} I_k(\varphi)$$
$$= \sum_{k=1}^{n} \sum_{e \in E_{P(k)}^{\mathrm{exc}}} \left[c_e(f_{P(k)}) - c_e(0) \right]$$
$$+ \underbrace{\sum_{k=1}^{n} \sum_{e \in E_{P(k)}^{\mathrm{com}}} \left[c_e \left(f_{P(k)} + \sum_{\substack{P:e \in P \\ P \neq P(k)}} f_P^k \right) - c_e \left(\sum_{\substack{P:e \in P \\ P \neq P(k)}} f_P^k \right) \right]}_{(\#)}.$$

We reorder $(\#)$ by first summing up over the edges. For a given edge e, consider the players P that use e, but not exclusively: $\{P : e \in E_P^{\mathrm{com}}\}$. Assume that these players become active in the following order $\{P_{i(1)}, \ldots, P_{i(\ell)}\}$, which is a subsequence of $\{P(k)\}_{k=1,\ldots,n}$. By E^c we denote the set of edges that are used by more than one player; by E^e we denote the set of edges that are used by exactly one player.

$$(\#) = \sum_{e \in E^c} \sum_{k:e \in E_{P(k)}^{\mathrm{com}}} \left[c_e \left(f_{P(k)} + \sum_{\substack{P:e \in P \\ P \neq P(k)}} f_P^k \right) - c_e \left(\sum_{\substack{P:e \in P \\ P \neq P(k)}} f_P^k \right) \right]$$

$$= \sum_{e \in E^c} \sum_{m=1}^{\ell} \left[c_e \left(\sum_{q=1}^{m} f_{P_{i(q)}} \right) - c_e \left(\sum_{q=1}^{m-1} f_{P_{i(q)}} \right) \right] \tag{2.33}$$

$$= \sum_{e \in E^c} \left[c_e \left(\sum_{q=1}^{\ell} f_{P_{i(q)}} \right) - c_e(0) \right]$$

$$= \sum_{e \in E^c} [c_e(f_e) - c_e(0)]. \tag{2.34}$$

(2.33) is true as in the direct sequence after the mth step, $f_{P_{i(q)}}^m = f_{P_{i(q)}}$ for $q \leq m$ and $f_{P_{i(q)}} = 0$ for $q > m$ holds. In addition, (2.34) holds because $f_e = \sum_{P:e \in P} f_P$.

By using that

$$\sum_{k=1}^{n} \sum_{e \in E_{P(k)}^{exc}} [c_e(f_{P(k)}) - c_e(0)] = \sum_{e \in E^e} \sum_{k:e \in E_{P(k)}^{exc}} [c_e(f_{P(k)}) - c_e(0)]$$

$$= \sum_{e \in E^e} [c_e(f_e) - c_e(0)],$$

we are able to conclude

$$I(\varphi) = \sum_{e \in E} [c_e(f_e) - c_e(0)]. \tag{2.35}$$

By Lemma 2.94 it holds that (2.35) is a an exact restricted potential function. As $c_e(0)$ is constant for all $e \in E$, we have that $\Pi(f) = \sum_{e \in E} c_e(f_e)$ is an exact restricted potential function, as well.   □

The next theorem follows immediately and gives an algorithmic approach for finding equilibria in arbitrary path player games.

Theorem 2.102. *In a path player game, solutions to the optimization problem*

$$\max \Pi(f) = \sum_{e \in E} c_e(f_e) \quad \text{subject to } f \in \mathbb{F} \tag{2.36}$$

are feasible equilibria.

Proof. By Lemma 2.92 on page 65, an optimal solution of (2.36) is a generalized equilibrium for the generalized path player game. By Theorem 2.90, each generalized equilibrium is a feasible equilibrium in the path player game.   □

The following existence statement is based on the previous theorem.

Theorem 2.103. *In path player games with continuous cost functions, feasible equilibria exist.*

Proof. In path player games, the set of feasible flows $\mathbb{F}$ is nonempty and compact by definition. Due to the continuous cost functions c_e, the restricted potential function $\Pi(f) = \sum_{e \in E} c_e(f_e)$ is continuous too and hence, optimal solutions of (2.36) do exist. The existence of feasible equilibria follows by Theorem 2.102.   □

This is an alternative proof of Theorem 2.31 on page 24. In Section 2.4.5, we propose a third approach for proving the existence of equilibria, again based on the existence of potential functions, see Theorem 2.118, page 84.

We conclude that by solving (2.36) we will find at least one equilibrium. On the other hand, not necessarily all equilibria may be found by this approach, which is a motivation to study other ways of computing equilibria; see Section 2.4.5.

2.4.3 An Ordinal Potential Function for Path Player Games

To construct an exact potential, in the previous section we reduced the strategy space to the feasible strategies. If we are considering the original path player game on the complete strategy space $f \in \mathbb{R}_+^{|\mathcal{P}|}$, the infeasibility penalty complicates the situation. By allowing infeasible strategies and thus infeasible sequences, we get problems with the cost of a sequence each time the sequence leaves the feasible strategy space $\mathbb{F}$ (see Example 2.88). Nevertheless, we are able to present an ordinal potential for path player games. For the proof, the next proposition is necessary.

Proposition 2.104. *Let* $f^x = (f_{-P}, x)$ *and* $f^z = (f_{-P}, z)$ *be two feasible flows that differ only in* f_P. *Then*

$$\sum_{e \in E} c_e(f_e^x) - \sum_{e \in E} c_e(f_e^z) = c_P(f^x) - c_P(f^z).$$

Proof.

$$\sum_{e \in E} c_e(f_e^x) - \sum_{e \in E} c_e(f_e^z)$$
$$= \sum_{e \in E} c_e(f_e^x) = \sum_{e \in P} c_e(f_e^x) \tag{2.37}$$
$$= c_P(f^x) - c_P(f^z).$$

Equation (2.37) is true as f^x and f^z are different only in f_P.   □

Theorem 2.105. *Path player games are ordinal potential games. An ordinal potential function is given by*

$$\Pi(f) = \begin{cases} \sum_{e \in E} c_e(f_e) & \text{if } \sum_{P \in \mathcal{P}} f_P \leq r \\ -M & \text{if } \sum_{P \in \mathcal{P}} f_P > r \end{cases}. \tag{2.38}$$

Proof. Consider any player P and any two flows (f_{-P}, x) and (f_{-P}, z). We distinguish four cases.

Case 1: (f_{-P}, x) infeasible, (f_{-P}, z) infeasible:

$$b_P(f_{-P}, x) - b_P(f_{-P}, z) = -M - (-M) = \Pi(f_{-P}, x) - \Pi(f_{-P}, z).$$

Case 2: (f_{-P}, x) infeasible, (f_{-P}, z) feasible:

$$b_P(f_{-P}, x) - b_P(f_{-P}, z) = -M - c_P(f_{-P}, z) < 0,$$

as $c_P(f) \geq 0$ for $f \in \mathbb{R}^n_+$ and as M is a sufficiently large number, at least $M > \sum_{e \in E} c_e(0)$. By the same argument and as $c_e(f_e) \geq 0$ for $f_e \in \mathbb{R}_+$,

$$\Pi(f_{-P}, x) - \Pi(f_{-P}, z) = -M - \sum_{e \in E} c_e(f_e) < 0,$$

holds.

Case 3: (f_{-P}, x) feasible, (f_{-P}, z) infeasible:

$$b_P(f_{-P}, x) - b_P(f_{-P}, z) = c_P(f_{-P}, x) - (-M) > 0,$$

as $c_P(f) \geq 0$ for $f \in \mathbb{R}^n_+$ and as M is a sufficiently large number, at least $M > \sum_{e \in E} c_e(0)$. By the same argument and as $c_e(f_e) \geq 0$ for $f_e \in \mathbb{R}_+$,

$$\Pi(f_{-P}, x) - \Pi(f_{-P}, z) = \sum_{e \in E} c_e(f_e) - (-M) > 0,$$

holds.

Case 4: (f_{-P}, x) feasible, (f_{-P}, z) feasible:

By Proposition 2.104 we get:

$$\Pi(f^x) - \Pi(f^z) = \sum_{e \in E} c_e(f^x_e) - \sum_{e \in E} c_e(f^z_e) =$$

$$c_P(f^x) - c_P(f^z) = b_P(f^x) - b_P(f^z).$$

Summarizing these four cases we conclude that

$$b_P(f_{-P}, x) - b_P(f_{-P}, z) > 0 \quad \Leftrightarrow \quad \Pi(f_{-P}, x) - \Pi(f_{-P}, z) > 0. \qquad \square$$

For two classes of path player games, namely for games on path-disjoint networks and for games with strictly increasing costs, we are able to give another description of an ordinal potential function.

Theorem 2.106. *In a path player game on a path-disjoint network or in a path player game where the cost functions c_e are strictly increasing, an ordinal potential function is given by*

$$\Pi(f) = \begin{cases} \sum_{P \in \mathcal{P}} c_P(f) & \text{if } \sum_{P \in \mathcal{P}} f_P \leq r \\ -M & \text{if } \sum_{P \in \mathcal{P}} f_P > r \end{cases}. \tag{2.39}$$

Proof. Consider any player P and any two flows (f_{-P}, x) and (f_{-P}, z). We distinguish the same four cases as in the proof of Theorem 2.105. The cases 1–3 are analogous to that proof. We have only to consider case 4, namely (f_{-P}, x) and (f_{-P}, z) feasible.

Path-disjoint network:

$$
\begin{aligned}
b_P(f_{-P}, x) &- b_P(f_{-P}, z) \\
&= c_P(f_{-P}, x) - c_P(f_{-P}, z) \\
&= c_P(x) - c_P(z) && (2.40) \\
&= c_P(x) - c_P(z) + \sum_{P_k \in \mathcal{P} \setminus \{P\}} c_{P_k}(f_{P_k}) - \sum_{P_k \in \mathcal{P} \setminus \{P\}} c_{P_k}(f_{P_k}) \\
&= \Pi(f_{-P}, x) - \Pi(f_{-P}, z). && (2.41)
\end{aligned}
$$

(2.40) and (2.41) hold due to the path-disjoint network.

Strictly increasing costs:
Assume

$$
\begin{aligned}
b_P(f_{-P}, x) - b_P(f_{-P}, z) &= c_P(f_{-P}, x) - c_P(f_{-P}, z) > 0 \\
&\Leftrightarrow x - z > 0 && (2.42) \\
&\Leftrightarrow \sum_{P_k \in \mathcal{P}} c_{P_k}(f_{-P}, x) - \sum_{P_k \in \mathcal{P}} c_{P_k}(f_{-P}, z) > 0 && (2.43) \\
&\Leftrightarrow \Pi(f_{-P}, x) - \Pi(f_{-P}, z) > 0.
\end{aligned}
$$

(2.42) is true because of strictly increasing costs c_e. Due to the same reason $c_P(f_{-P}, x) - c_P(f_{-P}, z) \geq 0 \; \forall \, P \in \mathcal{P}$ holds, which is used in (2.43). $\square$

Lemma 2.107. *In path player games on path-disjoint networks or in path player games with strictly increasing cost functions c_e, there exists an equilibrium which is nondominated (see Definition 2.68, page 50).*

Proof. Consider the equilibrium f^* which is given as a maximizer of the potential function (2.39). In Theorem 1.11 of [Kra05], it is proved that a flow maximizing the sum of the benefits over all players in a game is a nondominated flow. $\square$

In general it is not necessarily true that a maximizer f of a potential function is also nondominated. For instance see Example 2.77 on page 58, where each equilibrium is dominated. As the maximizer f is an equilibrium, it has to be dominated in this case. Note that maximizers exist for the potential functions presented in this work as they are continuous over the set of feasible flows. In addition, if they are defined for infeasible flows, they provide for these flows a negative potential value small enough such that the maximizer will lie within the feasible flows.

2.4.4 An Exact Potential for an Extended Benefit Function

We have already pointed out in Example 2.88 that the path player game in its original notation using the benefit function from Definition 2.4 is not an exact potential game. An extension of the standard benefit function is introduced, which allows the development of an exact potential for the path player game.

Definition 2.108. *In a path player game, the* extended benefit function *is given by*

$$b_P^{ext}(f) = \begin{cases} c_P(f) & \text{if } \sum_{P_k \in \mathcal{P}} f_{P_k} \leq r \\ -M + c_P(f) & \text{if } \sum_{P_k \in \mathcal{P}} f_{P_k} > r \end{cases}. \tag{2.44}$$

The extension concerns the second part of the function, the penalty for infeasible flows. The first part, which concerns feasible flows is not touched. The extended benefit function is in fact a realistic model of economic situations. A player gets punished for sending too much flow, but nevertheless, she receives the income created by the costs. For instance, consider a company that is producing more goods than it is allowed by a pollution regulation, and it gets punished thus. Nevertheless this company is selling all produced goods and obtains income from this.

Theorem 2.109. *The function*

$$\Pi(f) = \begin{cases} \sum_{e \in E} c_e(f_e) & \text{if } \sum_{P \in \mathcal{P}} f_P \leq r \\ -M + \sum_{e \in E} c_e(f_e) & \text{if } \sum_{P \in \mathcal{P}} f_P > r \end{cases} \tag{2.45}$$

is an exact potential function for any path player game with an extended benefit function.

Proof. Consider any player P and any two flows $f^x = (f_{-P}, x)$ and $f^z = (f_{-P}, z)$. We distinguish four cases. Proposition 2.104 is used in each case.

Case 1: (f^x) infeasible, (f^z) infeasible:

$$\Pi(f^x) - \Pi(f^z) = -M + \sum_{e \in E} c_e(f_e^x) - \left(-M + \sum_{e \in E} c_e(f_e^z) \right)$$
$$= -M + c_P(f^x) - (-M + c_P(f^z)) = b_P(f^x) - b_P(f^z).$$

Case 2: (f^x) infeasible, (f^z) feasible:

$$\Pi(f^x) - \Pi(f^z) = -M + \sum_{e \in E} c_e(f_e^x) - \sum_{e \in E} c_e(f_e^z)$$
$$= -M + c_P(f^x) - c_P(f^z) = b_P(f^x) - b_P(f^z).$$

Case 3: (f^x) feasible, (f^z) infeasible:

$$\Pi(f^x) - \Pi(f^z) = \sum_{e \in E} c_e(f_e^x) - \left(M + \sum_{e \in E} c_e(f_e^z) \right)$$
$$= c_P(f^x) - (M + c_P(f^z)) = b_P(f^x) - b_P(f^z).$$

Case 4: (f^x) feasible, (f^z) feasible:

$$\Pi(f^x) - \Pi(f^z) = \sum_{e \in E} c_e(f_e^x) - \sum_{e \in E} c_e(f_e^z)$$
$$= c_P(f^x) - c_P(f^z) = b_P(f^x) - b_P(f^z).$$

Summarizing, we conclude that

$$b_P(f^x) - b_P(f^z) = \Pi(f^x) - \Pi(f^z). \qquad \square$$

2.4.5 Computation of Equilibria by Improvement Sequences

One approach to obtain an equilibrium is to solve the optimization problem given in Theorem 2.102. The drawback is that in general we will not find all equilibria by this computation. In this section, another approach is presented. Dependent on the initial flows, the algorithm is in principle able to deliver each equilibrium in a path player game. A proper choice of the set of initial flows is in fact an open question and an analysis of the attraction region of the equilibria profiles is a topic of future research.

In Section 2.4.1 we introduced sequences and the finite improvement property (FIP). A finite improvement sequence terminates with an equilibrium, which is a motivation to study this approach. Due to the infinite number of strategies, path player games do not satisfy FIP and we have to look for alternatives. We investigate two different approaches in this section. The first one uses *best-reply improvement sequences* which are proposed in [Mil96]. We show in this section that in path player games, best-reply improvement sequences are in general not finite. Nevertheless, we present classes of path player games, where best-reply sequences are finite and thus end with an equilibrium. The second approach analyzes ε-improvement sequences, which yield *approximate equilibria*; see [MS96].

First, we consider the finite best-reply property, which is a greedy approach. At each step the active player chooses a best reply as a new strategy such that the improvement of the active player's benefit is maximized.

Definition 2.110. *A best-reply sequence is a strategy sequence according to Definition 2.82, page 61, where in each step the active player shifts to a best-reply strategy with respect to the strategies of his competitors:*

$$f_{P(k)}^k \in f_{P(k)}^{\max} \left(f_{-P(k)}^{k-1} \right)$$

holds for all $k = 1, \ldots, N$.

A best-reply sequence is called a best-reply improvement sequence *if a player will only become active if she can obtain a strict improvement, that means if*

$$f_{P(k)}^{k-1} \notin f_{P(k)}^{\max} \left(f_{-P(k)}^{k-1} \right)$$

holds for all $k = 1, \ldots, N$.

A game satisfies the finite best-reply property *(FBRP)* if every best-reply improvement sequence is finite.*

The finite improvement property (FIP) implies the finite best-reply property but the reverse is not true. Unfortunately, best-reply improvement sequences may also be infinite in path player games; that is, FBRP is not satisfied in general. This fact is illustrated by the following example.

Example 2.111. Consider a path player game with two players and a flow rate $r = 1$. Each of the two paths consists of two edges. One is owned exclusively by the player of this path; the other one is shared with the second player (see Figure 2.40).

Let f_1, f_2 be the flows of player 1 and 2. The cost of the exclusively used edges are given by $c_{e_1}(f_1) = -f_1^2 + 1$ for player 1 and $c_{e_2}(f_2) = -f_2^2 + 1$ for player 2. The cost of the commonly used edge is $c_{e_3}(f) = -(f_1 + f_2 - 1)^2 + 1$. Thus, we get the cost of the paths as $c_1(f) = -f_1^2 - (f_1 + f_2 - 1)^2 + 2$ and $c_2(f) = -f_2^2 - (f_1 + f_2 - 1)^2 + 2$.

Given a fixed f_2, the first player will choose $f_1 = 1/2 - f_2/2$ as the best reply, whereas player 2 will choose $f_2 = 1/2 - f_1/2$ for a given f_1. This best-reply mapping has a fixed point at $f^* = (1/3, 1/3)$ which is also the unique equilibrium of the game. See Figure 2.41 for an illustration of the set of feasible solutions and the best-reply strategies. The equilibrium f^* can only be reached by the best-reply mapping if the initial flow f^0 has either $f_1^0 = 1/3$ or $f_2^0 = 1/3$ or both. Any other start point will create an infinite best-reply sequence; see, for example, Figure 2.42. Nevertheless, in the next theorem we present a class of path player games that satisfy FBRP.

Theorem 2.112. *Consider a path player game where for all players $P \in \mathcal{P}$ and for all $f_{-P} \in \mathbb{R}_+^{|\mathcal{P}|-1}$ the best reaction sets satisfy*

$$f_P^{\max} \subseteq \{0, d_P\}. \tag{2.46}$$

This game satisfies FBRP.

Proof. Consider a path player game with n players $P = P_1, \ldots, P_n$ and a best-reply improvement sequence $\varphi = (f^0, f^1, \ldots, f^k, \ldots)$. Let $d^k = r - \sum_{P \in \mathcal{P}} f_P^k$ be the *free flow rate at step k*.

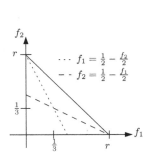

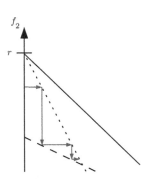

Fig. 2.40.
Game network.

Fig. 2.41.
Best-reply strategies.

Fig. 2.42. Best-reply
improvement sequence.

We introduce a fictitious player $P = P_{n+1}$ who is "sending" the free flow rate as her own strategy; that is, $f^k_{P_{n+1}} = d^k$ holds for all $k \geq 0$. For each player $P = P_1, \ldots, P_n$ and each step $k \geq 0$, we define the set $A^k_P \subseteq \{P_1, \ldots, P_{n+1}\}$ that is a subset of the set of players (including the fictitious one). In addition, for each $k > 0$, the sets $A^k_{P_1}, \ldots, A^k_{P_{n+1}}$ have to be a partition of the set $\{P_1, \ldots, P_{n+1}\}$; that is,

$$\bigcup_{P=P_1,\ldots,P_{n+1}} A^k_P = \{P_1, \ldots, P_{n+1}\}, \tag{2.47}$$

$$A^k_P \cap A^k_{P'} = \emptyset \quad \forall \, P \neq P', \; P, P' \in \{P_1, \ldots, P_{n+1}\} \tag{2.48}$$

has to hold for all $k > 0$.

In the following, we use these sets to show that in each step k, each initial flow f^0_P contributes to the flow of exactly one player.

Claim: For each $k > 0$ there is a partition of A^k_P, $P = P_1, \ldots, P_{n+1}$ (satisfying (2.47) and (2.48)), such that

$$f^k_P = \sum_{P_j \in A^k_P} f^0_{P_j}$$

holds for all $P = P_1, \ldots, P_{n+1}$.

We prove the claim by induction: $k = 0$.

Set $A^0_P = \{P\} \; \forall \, P = P_1, \ldots, P_{n+1}$. It follows that

$$\sum_{P_j \in A^0_P} f^0_{P_j} = \sum_{P_j \in \{P\}} f^0_{P_j} = f^0_P.$$

$(k-1) \to k$.

Assume the claim is true for step $k-1$. Consider the active player $P(k)$, who is changing his flow from $f_{P(k)}^{k-1}$ to $f_{P(k)}^{k}$. By (2.46), one of the two following cases has to be satisfied.

(i) $f_{P(k)}^{k} = 0$.

(ii) $f_{P(k)}^{k} = f_{P(k)}^{k-1} + d^{k-1} = f_{P(k)}^{k-1} + f_{P_{n+1}}^{k-1}$.

In the second case, we use that the fictitious player represents the free flow rate. Automatically, his flow $f_{P_{n+1}}^{k}$ is also changed by the move of the active player. According to the cases (i) and (ii) we have

(i) $f_{P_{n+1}}^{k} = f_{P_{n+1}}^{k-1} + f_{P(k)}^{k-1}$.

(ii) $f_{P_{n+1}}^{k} = 0$.

Furthermore, $f_P^k = f_P^{k-1}$ holds for all $P \notin \{P(k), P_{n+1}\}$.

For case (i), set $A_{P(k)}^k = \emptyset$, $A_{P_{n+1}}^k = A_{P_{n+1}}^{k-1} \cup A_{P(k)}^{k-1}$, and $A_P^k = A_P^{k-1} \; \forall \, P \notin \{P(k), P_{n+1}\}$. This is a partition, as (2.47) and (2.48) are satisfied. Furthermore, we have

$$\sum_{P_j \in A_{P(k)}^k} f_{P_j}^0 = \sum_{P_j \in \emptyset} f_{P_j}^0 = 0,$$

$$\sum_{P_j \in A_{P_{n+1}}^k} f_{P_j}^0 = \sum_{P_j \in A_{P_{n+1}}^{k-1}} f_{P_j}^0 + \sum_{P_j \in A_{P(k)}^{k-1}} f_{P_j}^0 = f_{P_{n+1}}^{k-1} + f_{P(k)}^{k-1}, \qquad (2.49)$$

and for all $P \notin \{P(k), P_{n+1}\}$:

$$\sum_{P_j \in A_P^k} f_{P_j}^0 = \sum_{P_j \in A_P^{k-1}} f_{P_j}^0 = f_P^{k-1} = f_P^k.$$

Note that in (2.49) we use that $A_{P_{n+1}}^{k-1}$ and $A_{P(k)}^{k-1}$ are disjoint, because by the induction hypothesis (2.48) is satisfied.

For case (ii), set $A_{P(k)}^k = A_{P(k)}^{k-1} \cup A_{P_{n+1}}^{k-1}$, $A_{P_{n+1}}^k = \emptyset$, and $A_P^k = A_P^{k-1} \; \forall \, P \notin \{P(k), P_{n+1}\}$. This is a partition, as (2.47) and (2.48) are satisfied. Furthermore, we have

$$\sum_{P_j \in A_{P(k)}^k} f_{P_j}^0 = \sum_{P_j \in A_{P(k)}^{k-1}} f_{P_j}^0 + \sum_{P_j \in A_{P_{n+1}}^{k-1}} f_{P_j}^0 = f_{P(k)}^{k-1} + f_{P_{n+1}}^{k-1}, \qquad (2.50)$$

$$\sum_{P_j \in A_{P_{n+1}}^k} f_{P_j}^0 = \sum_{P_j \in \emptyset} = 0,$$

and for all $P \notin \{P(k), P_{n+1}\}$:

$$\sum_{P_j \in A_P^k} f_{P_j}^0 = \sum_{P_j \in A_P^{k-1}} f_{P_j}^0 = f_P^{k-1} = f_P^k.$$

In (2.50) it is used that $A_{P_{n+1}}^{k-1}$ and $A_{P(k)}^{k-1}$ are disjoint, because by the induction hypothesis (2.48) is satisfied.

This last case finishes the proof of the claim.

There are a finite number of possibilities to partition $\{P_1, \ldots, P_{n+1}\}$ into sets A_P^k. Thus, the number of different flows f^k that can be obtained by a best-reply improvement sequence is finite. Furthermore, in a best-reply improvement sequence, no flow f is visited twice. This is true as the path player game is a potential game and thus, for a best-reply improvement sequence, a strict improvement of the potential function $\Pi(f)$ is required in each step. By this argument, a cycle does not exist in any best-reply improvement path, and in addition the number of different flows f^k is finite, therefore each best-reply improvement path is finite. $\square$

Corollary 2.113. *The following classes of path player games satisfy FBRP.*

- *PPG with cost functions c_e that are linear in f_P*
- *PPG with cost functions c_e that are strictly increasing in f_P*
- *PPG with cost functions c_e that are strictly convex in f_e*

The corollary follows immediately from Theorem 2.112, as the described cost functions attain their maximum on the boundary of a compact interval. As terminal flows of maximal best-reply improvement sequences are equilibria, we can determine equilibria for the presented classes of path player games within a finite number of steps.

For general path player games, Example 2.111 illustrates that best-reply improvement sequences may be infinite. Therefore, we need to consider other ways for the determination of Nash equilibria. In the next approach, we determine *approximate equilibria*, which are described in [MS96].

Definition 2.114. *For $\varepsilon > 0$, a sequence $\varphi = (f^0, \ldots, f^k, \ldots)$ is an ε-improvement sequence if for all $k \geq 1$ it holds*

$$b_{P(k)}(f_k) > b_{P(k)}(f_{k-1}) + \varepsilon.$$

A game Γ satisfies the approximate finite improvement property (AFIP) *if for all $\varepsilon > 0$ every ε-improvement sequences is finite.*

A maximal finite ε-improvement sequence yields a terminal flow, where no player is able to improve the benefit by more than ε. We define the following.

Definition 2.115. *For $\varepsilon > 0$, a flow f^ε that satisfies $\forall\, P \in \mathcal{P}$ and $\forall\, f_P \geq 0$,*

$$b_P(f_{-P}^\varepsilon, f_P^\varepsilon) \geq b_P(f_{-P}^\varepsilon, f_P) - \varepsilon,$$

is called an ε-equilibrium.

Clearly, if an ε-improvement sequence is finite, it terminates with an ε-equilibrium. Note that ε-equilibria may be infeasible. See Lemma 2.32, page 28 for a characterization of infeasible equilibria in path player games. However, the only possibility to obtain an infeasible ε-equilibrium f^* as the terminal flow of an ε-improvement sequence is that the sequence has started with f^*. In this case, the sequence has length zero. We neglect this degenerated case and concentrate on feasible ε-equilibria.

Theorem 2.116. *The path player game satisfies AFIP.*

Proof. Consider an ε-improvement sequence φ. First assume the initial flow f^0 of φ is infeasible. Then, two cases may take place: Either no player is able to improve his benefit; that means f^0 is an equilibrium and φ terminates with $f^N = f^0$ and $l(\varphi) = 0$ or, there is at least one player who is able to improve her own benefit by creating a feasible flow within one step. Then f^1 of φ is feasible.

For an improvement sequence it holds that if f^k is feasible, the subsequent flow f^{k+1} is feasible, too. Thus it is sufficient, for the remaining proof, to consider the set of feasible flows $\mathbb{F}$ as ε-improvement sequences with $l(\varphi) > 0$ will either start in $\mathbb{F}$ or jump into $\mathbb{F}$ in the first step.

The benefit functions in path player games are bounded for $f \in \mathbb{F}$, as $b_P(f) = c_P(f)$ is continuous over the closed set $\mathbb{F}$. Thus, the restricted potential function $\Pi(f)$, which exists according to Theorem 2.96 (page 68), is bounded as well. As an ε-improvement sequence increases the restricted potential function values by at least an $\varepsilon > 0$ in each step, each ε-improvement sequence has to be finite.   □

Note that AFIP is also satisfied for any exact potential game with bounded benefit functions, which is a result of [MS96]. If we increase the precision for an ε-improvement sequence by decreasing ε, we obtain a sequence of feasible ε-equilibria. By the following lemma, accumulation points of that sequence are equilibria, thus ε-improvement sequences can be used for the computation of equilibria with a given precision.

Lemma 2.117. *Consider a path player game and a sequence of feasible ε-equilibria $f^*(\varepsilon)$ that is given for $\varepsilon \to 0$. Any accumulation point f^* is a feasible equilibrium of the path player game.*

Proof. First, note that feasible ε-equilibria exist for each $\varepsilon > 0$ as path player games satisfy AFIP. Furthermore, by the feasibility of ε-equilibria, the sequence $\{f^*(\varepsilon)\}_{\varepsilon \to 0}$ is bounded. Hence, by the Bolzano–Weierstrass theorem, it has an accumulation point; that is, we can find a subsequence $f^*(\bar{\varepsilon}^k)$ that converges to f^* for $k \to \infty$. Furthermore, $\bar{\varepsilon}^k \overset{k \to \infty}{\longrightarrow} 0$ because this holds for each subsequence of the original sequence.

By the definition of ε-equilibria it has to hold for each $\varepsilon > 0$ that

$$b_P(f^*_{-P}(\varepsilon), f^*_P(\varepsilon)) - b_P(f^*_{-P}(\varepsilon), f_P) \geq \varepsilon \quad \forall P \in \mathcal{P}, \forall f_P \geq 0. \qquad (2.51)$$

Now consider the limit of (2.51):

$$\lim_{\varepsilon \to 0} b_P(f^*_{-P}(\varepsilon), f^*_P(\varepsilon)) - \lim_{\varepsilon \to 0} b_P(f^*_{-P}(\varepsilon), f_P) \geq \lim_{\varepsilon \to 0} \varepsilon \quad \forall\, P \in \mathcal{P}, \forall f_P \geq 0.$$

Because we assume continuous functions c_e, the benefit, which is given as $b_P(f) = c_P(f)$ for $f \in \mathbb{F}$, is upper semi-continuous for $f \in \mathbb{F}$ and we rewrite:

$$b_P \left(\lim_{\varepsilon \to 0} f^*_{-P}(\varepsilon), \lim_{\varepsilon \to 0} f^*_P(\varepsilon) \right) - b_P \left(\lim_{\varepsilon \to 0} f^*_{-P}(\varepsilon), f_P \right) \geq 0 \quad \forall\, P \in \mathcal{P}, \forall f_P \geq 0$$

$$\Leftrightarrow b_P \left(f^*_{-P}, f^*_P \right) - b_P \left(f^*_{-P}, f_P \right) \geq 0 \quad \forall\, P \in \mathcal{P}, \forall f_P \geq 0\ ;$$

thath is, f^* is a feasible equilibrium. $\square$

The following theorem is obtained directly from the previous lemma.

Theorem 2.118. *In path player games with continuous cost functions c_e, feasible equilibria exist.*

The above argument provides an alternative proof of the existence of pure strategy equilibria in path player games. Note that the existence of equilibria is also a direct consequence of Theorem 2.102, where it is shown that the solutions of a particular optimization problem are equilibria. See Theorem 2.103 on page 73 for the corresponding existence statement. These two approaches of proving the existence of equilibria are motivated by the existence of potential functions for that kind of games. In addition, there is a first existence proof that uses a fixed-point argument; see Theorem 2.31 on page 24.

Note that Monderer and Shapley have provided in [MS96] a sufficient condition for the existence of equilibria in infinite potential games. However, as path player games have noncontinuous benefit functions, they do not satisfy the requirements in the reference.

3

Games on Polyhedra: A Generalization

Given a polyhedron $S \subseteq \mathbb{R}^n$ and payoff functions $c_i : S \to \mathbb{R}$, one for each player $i = 1, \ldots, n$, find a coordinate x_i for each player such that $x = (x_1, \ldots, x_n)$ lies within S and the payoff of each player is maximized with respect to the chosen coordinates of the competitors x_{-i}.

3.1 Introduction

In this chapter, a generalization of path player games is investigated. *Games on polyhedra* is a class of infinite games, in which the set of feasible strategies of a player depends on the strategies chosen by other players. We present results for special instances of payoff functions concerning equilibria and their existence and dominated solutions. Moreover, we show how such games can be transformed to equivalent games on a hypercuboid and investigate the existence of a potential.

The game on a polyhedron is played by a finite number of players, $i = 1, \ldots, n$. Each of these players chooses a strategy x_i such that the vector of strategies $x = (x_i)_{i=1,\ldots,n}$, which we call the *solution*, lies within a polyhedron $S(A, b) = \{x : Ax \leq b\}$, where $A = (a_{ji}) \in \mathbb{R}^{m \times n}$ and $b = (b_j) \in \mathbb{R}^m$ are the parameters of the game. By this restriction, the set of strategies a player is allowed to choose from depends on the strategies the other players choose. This is the crucial property of *generalized Nash equilibrium (GNE) games*, a class of games to which the games on polyhedra belong. We present a short review of the literature about GNE games at the end of this section. The *payoff* is represented by a continuous mapping $c : S(A, b) \to \mathbb{R}^n$, where $c_i(x)$ denotes the payoff of player i. The vector $x_{-i} \in \mathbb{R}^{n-1}$ is given by $x_{-i} = (x_1, \ldots, x_{i-1}, x_{i+1}, \ldots, x_n)$.

Definition 3.1. *For a given x_{-i} the set of* feasible strategies *for the ith player is denoted by*

$$S_i(x_{-i}) = \{x_i : A(x_1, \ldots, x_i, \ldots, x_n)^T \leq b\}.$$

S. Schwarze, *Path Player Games*, DOI 10.1007/978-0-387-77928-7_3,
© Springer Science+Business Media, LLC 2009

We say the vector x_{-i} is feasible *if $S_i(x_{-i})$ is nonempty, and* infeasible *otherwise.*

A solution x is feasible, *if $x \in S(A, b)$ and* infeasible *otherwise.*

Example 3.2. Consider a path player game with no security limit; that is, $\omega_P = 0 \ \forall \ P \in \mathcal{P}$ (see page 12). If we consider only feasible flows and neglect infeasible ones (see generalized path player games, Section 2.4.2 on page 64), the game can be modeled as a game on a polyhedron $S(A, b)$. Let the number of players be given by $n = |\mathcal{P}|$. The $(n+1) \times n$-matrix A and the n-dimensional vector b describe the game on the polyhedron:

$$A = \begin{pmatrix} 1 & 1 \ldots & 1 \\ -1 & 0 \ldots & 0 \\ 0 & -1 \ldots & 0 \\ \vdots & \vdots \ddots & \vdots \\ 0 & 0 \ldots & -1 \end{pmatrix} \quad b = \begin{pmatrix} r \\ 0 \\ \vdots \\ 0 \end{pmatrix}.$$

See Figure 3.1 for the visualization of the polyhedron $S(A, b)$ for a path player game in the two-player version.

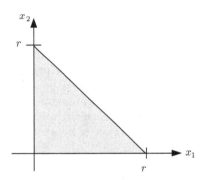

Fig. 3.1. Polyhedron for a path player game with two players.

If we have in contrast a path player game with security limits $\omega_P > 0$, the representation as a game on a polyhedron is not so simple. We suggest two possibilities.

– Game on polyhedron with noncontinuous payoffs. A path player game with general security limits $\omega_P > 0$ can be modeled as a game on a polyhedron if noncontinuous payoffs are allowed.

  Consider a player P with cost $\bar{c}_P(f)$ and let P be the ith player. In the game on a polyhedron the payoff of player P is modeled by

$$c_i(f) = \begin{cases} \bar{c}_P(f) & \text{if } f_P \geq \omega_P \\ \kappa_P & \text{if } f_P < \omega_P \end{cases}.$$

This approach has the main disadvantage that results of the following sections (e.g., the existence of potential functions) cannot be applied within this approach, because they do not hold for noncontinuous payoff functions.

– Security limit as hard constraint. If we consider the inequalities $f_P \geq \omega_P$ as constraints of the game (i.e., we do not allow a security payment κ_P), we can model the path player game as a game on a polyhedron. By adding the constraints $f_P \geq \omega_P$ to the description of the $S(A, b)$ we obtain:

$$A = \begin{pmatrix} 1 & 1 & \dots & 1 \\ -1 & 0 & \dots & 0 \\ 0 & -1 & \dots & 0 \\ \vdots & \vdots & \ddots & \vdots \\ 0 & 0 & \dots & -1 \end{pmatrix} \qquad b = \begin{pmatrix} r \\ -\omega_1 \\ -\omega_2 \\ \vdots \\ -\omega_n \end{pmatrix}.$$

This approach is interesting in cases where κ_P is used as a penalty for infeasible flows. For instance, this is true in line planning games, which are described in Chapter 4. Although we have a different structure of security limits in line planning games, they can be modeled as games on polyhedra without problems. See the referred chapter for details.

Example 3.3. Consider a game on a polyhedron with

$$A = \begin{pmatrix} 1 & 1 \\ -1 & 1 \\ -1 & -1 \\ -1 & 0 \\ 0 & -1 \end{pmatrix} \qquad b = \begin{pmatrix} 3 \\ 2 \\ -1 \\ 0 \\ 0 \end{pmatrix},$$

as illustrated in Figure 3.2. For a given $x_{-1} = (1)$ we have

$$S_1(1) = \{x_1 : x_1 \leq 2, -x_1 \leq 0, -x_1 \leq 1\} = \{x_1 : 0 \leq x_1 \leq 2\};$$

that is, $x_{-1} = (1)$ is feasible. Also solution $x_{-1} = (2.5)$ is feasible with

$$S_1(2.5) = \{x_1 : x_1 \leq 0.5, -x_1 \leq -0.5, -x_1 \leq 1.5, -x_1 \leq 0\} = \{0.5\}.$$

The vector $x_{-1} = (3)$ is infeasible because it yields $S_1(3) = \emptyset$.

In the game on a polyhedron, the players want to maximize their payoffs with respect to the decisions of their competitors. The equilibrium in this game is defined in the following way.

Definition 3.4. *A feasible solution x^* is a* generalized equilibrium *(in short: equilibrium) if and only if for all players $i = 1, \dots, n$ it holds that*

$$c_i(x^*_{-i}, x^*_i) \geq c_i(x^*_{-i}, x_i) \quad \forall \, x_i \in S_i(x^*_{-i}). \tag{3.1}$$

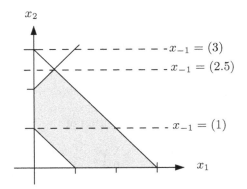

Fig. 3.2. Example 3.3.

Games on polyhedra are instances of the class of generalized Nash equilibria (GNE) games. In GNE games, a player i chooses from a strategy set $S_i(x_{-i})$ that is dependent on the strategies x_{-i} of the competitors. S denotes the set of feasible solutions; that is, $S = \{x : x_i \in S_i(x_{-i}) \; \forall \; i = 1, \ldots, n\}$ (in games on polyhedra, S is a polyhedron). For these types of games we find different names in the literature: GNE games [Har91], social equilibria games [Deb52], and abstract economies [AD54].

In terms of practical relevance of GNE games, Fukushima and Pang [FP05] applied the GNE theory to the *multi–leader–follower problem*. The strategies of the leaders are independent of the other players, whereas the follower's problem is a GNE game that is restricted by the strategies of the leaders. Harker [Har91] proposes the following example for that application. A government sets a maximum pollution level for a common used resource, for example, a body of water.

Another application is given for *games on squares* (see [Owe95]) where we have two players and the polyhedron is given as a square. An example for a game on a square is the *game of timing*, where the players have to choose a point in time for an activity (e.g., selling a good) within a given time frame. A player wins if he becomes active later than the opponent. As S is a square, the two players do not influence each other's strategy sets, such that the game could be called a degenerated GNE game. In [Owe95], mixed strategies are considered for investigating games on squares. In addition, for concave payoffs it is shown that equilibria in pure strategies exist.

Concerning the existence of equilibria, the generalized definition of the strategy sets makes the problem difficult. If there are no assumptions on the nature of the strategy sets and on how the players affect each other's strategy sets, then it is already a difficult problem even to find a feasible solution; that is, a solution that lies in the strategy sets of all players (see [FP06]). Thus, the existence of generalized equilibria is an often-discussed question and various existence theorems have been proposed in that area. Debreu [Deb52] and Arrow and Debreu [AD54] showed the existence of generalized equilibria on a compact and convex set S and for continuous and quasi-concave payoffs by using a fixed-point argument. Rosen [Ros65] has also studied instances of GNE games where S is a convex and compact set. Furthermore, he restricted his research to concave payoffs. He proved the existence of equilibria for this game by using a fixed-point argument. Even more, for particular payoff functions he showed uniqueness of the equilibria. Computation of these problems using gradient methods was discussed. Harker [Har91] exploited the relation of the GNE problem and *quasi-variational inequalities (QVI)*. For concave payoffs (in a maximization problem), GNE games can be described as QVI problems, and thus existence results can be transferred from QVI problems to GNE games. Harker used a result from Chan and Pang [CP82] stated for QVI problems, to prove the existence of GNE for continuous and concave payoffs c_i, provided that the mapping that describes $S_i(x_{-i})$ is nonempty, continuous, closed, and convex. Finally, in Section 2.4.2 we introduced the generalized path player game which can be described as a GNE game. In particular, it is an instance of the class of games on polyhedra. In Theorem 2.31, the existence of equilibria even in the case of a special type of noncontinuous payoff is proved.

In terms of uniqueness of equilibria, it cannot be guaranteed in general, as we can find examples where multiple equilibria appear quite often, for example, in generalized path player games; see Section 2.4.2.

Concerning computational questions, Facchinei and Pang [FP06] propose a penalty method for solving the GNE problem. In their approach, the strategy sets of the players are not restricted any more, but infeasible strategies get punished with an additional negative term in the payoff, related to the Langrangian relaxation approach. Note that either no or all players get punished, because the complete solution x is always feasible or infeasible. The new unrestricted problem can be solved using techniques from nonlinear optimization, for instance, Karush–Kuhn–Tucker or Fritz–John points. Fukushima and Pang [FP05] introduce an iterated penalty method where a sequence of variational inequality problems are solved to obtain a solution of the QVI problem. To ensure the relation to the GNE game, concavity of the payoffs (for maximization) is required here. A penalty approach is also used in path player games, introduced in Chapter 2, where infeasible strategies get punished with a constant negative payment $-M$.

Introducing games on polyhedra in this chapter, we describe an instance of the GNE games not discussed in the literature yet (apart from the games on a square, which are not GNE games in the classical sense), as to our knowledge no one has restricted the set S to be a polyhedron. In contrast to [Deb52, AD54, Ros65] and the approaches using QVI, we use for the beginning no restrictions on the payoff functions (apart from continuity). Later on, we restrict the payoffs, but by different assumptions from those in the mentioned references. As a result, existence of equilibria in games on polyhedra is not given in general. Nevertheless, existence is shown for linear payoffs, for strictly increasing payoffs, if the path player game property (see Definition 3.35) holds. Also, we describe algorithmic approaches for the computation of equilibria for special instances of games on polyhedra. Potential functions are studied and their existence is proved for two classes of games on polyhedra.

The order of the next sections is as follows. In Section 3.2, we discuss the existence of equilibria and their relation to nondominated solutions. In the same section, we introduce complete characterizations of the set of equilibria for two special instances of games on polyhedra. The game on a polyhedron is extended to a game on a hypercuboid in Section 3.3, which results in a game with independent strategy sets. In Section 3.4, potential functions for games on polyhedra as well as for games on hypercuboids are developed. These potential functions deliver additional insights on existence and computation of equilibria in the considered games.

3.2 Equilibria and Nondominated Solutions

Equilibria in games on polyhedra were already given by Definition 3.4. We have assumed continuous payoffs in these games, which is not sufficient for the existence of equilibria, as the following example illustrates.

Example 3.5. Consider a game on a polyhedron $S(A, b)$ with

$$
A = \begin{pmatrix} 1 & 1 \\ 1 & 0 \\ 0 & 1 \\ -1 & 0 \\ 0 & -1 \end{pmatrix} \quad b = \begin{pmatrix} 1.9 \\ 1 \\ 1 \\ 0 \\ 0 \end{pmatrix},
$$

as illustrated in Figure 3.3. The payoff is given by: $c_1(x) = -(x_1 - x_2)^2$ and $c_2(x) = (x_1 - x_2)^2$. Note that c_1 is concave in x_1 and c_2 convex in x_2. As the players want to maximize their payoffs, player 1 will try to minimize the distance between x_1 and x_2, whereas player 2 will try the opposite.

This game has no equilibrium. Suppose there were an interior point of S that was an equilibrium. Then either $x_1 = x_2$ or $x_1 \neq x_2$ would hold. In the first case player 2 could improve his payoff by increasing (or decreasing) x_2 by an ε sufficiently small. In the latter case player 1 is able to improve $c_1(x)$ by setting

$$
\bar{x}_1 = \begin{cases} x_1 + \varepsilon & \text{if } x_2 > x_1 \\ x_1 - \varepsilon & \text{if } x_2 < x_1 \end{cases},
$$

with ε sufficiently small. Hence, no interior point is an equilibrium. Also, no point on the boundary of S can be an equilibrium point:

For $\{x = (0, x_2) : 0 \leq x_2 < 1\}$ player 2 can improve his payoff by choosing $\bar{x}_2 = 1$.

For $\{x = (x_1, 1) : 0 \leq x_1 < 0.9\}$ player 1 improves with $\bar{x}_1 = 0.9$.

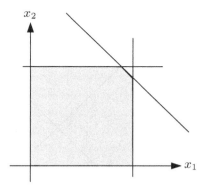

Fig. 3.3. No equilibrium exists.

For $\{x = (x_1, x_2) : x_1 + x_2 = 1.9, x_1, x_2 \geq 0.9\}$ player 2 improves with $\bar{x}_2 = 0$.
For $\{x = (1, x_2) : 0 < x_2 < 0.9\}$ player 2 improves with $\bar{x}_2 = 0$.
For $\{x = (x_1, 0) : 0 < x_1 \leq 1\}$ player 1 improves with $\bar{x}_1 = 0$.

Nevertheless, equilibria may exist for the same $c_1(x)$ and $c_2(x)$ in other polyhedra. For instance, consider polyhedron $S(\bar{A}, \bar{b})$, with

$$\bar{A} = \begin{pmatrix} 1 & 1 \\ 1 & -1 \\ -1 & 1 \\ -1 & 0 \\ 0 & -1 \end{pmatrix} \qquad \bar{b} = \begin{pmatrix} 2 \\ 1 \\ 1 \\ 0 \\ 0 \end{pmatrix},$$

as illustrated in Figure 3.4. Here $x^* = (0.5, 1.5)$ is an equilibrium, provided that the payoff functions are c_1, c_2 as given before. Then we have for player 1: $S_1(1.5) = \{0.5\}$ and for player 2: $S_2(0.5) = [0, 1.5]$, where

$$\max_{x_2 \in [0, 1.5]} (0.5 - x_2)^2 = 1$$

is obtained for $x_2 = 1.5$.

3.2.1 Existence and Characterization of Equilibria for Linear Payoffs

To investigate the existence of equilibria, we consider only compact polyhedra, as for noncompact polyhedra we can easily construct instances where equilibria do not exist. Note that a polyhedron $S(A, b) = \{x : Ax \leq b\}$ is closed by definition, such that it will be sufficient to require bounded polyhedra $S(A, b)$.

Consider games on compact polyhedra and with linear payoffs $c_i(x) = \sum_{j=1}^{n} \alpha_{ij} x_j$, $\alpha_{ij} \in \mathbb{R}$. We neglect the trivial case, where $\alpha_{ii} = 0$ holds for all

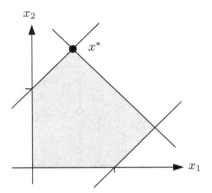

Fig. 3.4. An equilibrium exists.

$i = 1, \ldots, n$, as in this case every feasible solution is an equilibrium. For these games, we present a necessary and sufficient condition of equilibria. Let

$$\mathrm{sgn}(a) = \begin{cases} 1 & \text{if } a > 0 \\ -1 & \text{if } a < 0 \\ 0 & \text{if } a = 0 \end{cases}.$$

Theorem 3.6. *Consider a game on a compact polyhedron $S(A, b)$, where the payoff functions are linear (i.e., $c_i(x) = \sum_{j=1}^n \alpha_{ij} x_j$, $\alpha_{ij} \in \mathbb{R}$). We assume $\exists\, i : \alpha_{ii} \neq 0$. Furthermore, consider the following linear program,*

$$\max \sum_{i=1}^n \mathrm{sgn}(\alpha_{ii}) \lambda_i x_i \quad \text{subject to } x \in S(A, b). \tag{3.2}$$

In this game, the following hold.

(a) If the solution x^* is an equilibrium, then there exists a vector $\lambda \in \mathbb{R}_+^n$ that satisfies $\lambda_i > 0\ \forall\, i$ with $\alpha_{ii} \neq 0$, such that x^* solves (3.2).
(b) For all $\lambda \in \mathbb{R}_+^n$ that satisfy $\lambda_i > 0\ \forall\, i$ with $\alpha_{ii} \neq 0$, the optimal solution of (3.2) is an equilibrium.

Proof.
Part (a) x^* equilibrium $\Rightarrow$ $\exists\, \lambda \in \mathbb{R}_+^n : \lambda_i > 0\ \forall\, i$ with $\alpha_{ii} \neq 0$, such that x^* solves (3.2).
As x^* is an equilibrium, it has to lie on the boundary of $S(A, b)$: Suppose not; that is, x^* is an interior point. Then there exists a player i and an improvement direction v_i, (with v_i being the ith unit vector) such that $x^* + \delta v_i$ is in $S(A, b)$ for some positive and sufficiently small δ.

It follows that there is a constraint j such that it holds:

$$\sum_{i=1}^n a_{ji} x_i^* = b_j. \tag{3.3}$$

We fix:

$$\lambda_i = \begin{cases} |a_{ji}| & \text{if } \alpha_{ii} \neq 0 \ \wedge \ \text{sgn}(\alpha_{ii}) = \text{sgn}(a_{ji}) \\ \varepsilon & \text{if } \alpha_{ii} \neq 0 \ \wedge \ \text{sgn}(\alpha_{ii}) \neq \text{sgn}(a_{ji}) \ , \\ 0 & \text{if } \alpha_{ii} = 0 \end{cases} \tag{3.4}$$

with $\varepsilon > 0$, sufficiently small. Furthermore, we define the two sets:

$$G_1 = \{i : \alpha_{ii} \neq 0 \ \wedge \ \text{sgn}(\alpha_{ii}) = \text{sgn}(a_{ji})\} \, ,$$
$$G_2 = \{i : \alpha_{ii} \neq 0 \ \wedge \ \text{sgn}(\alpha_{ii}) \neq \text{sgn}(a_{ji})\} \, .$$

Then we get:

$$\sum_{i=1}^{n} \text{sgn}(\alpha_{ii})\lambda_i x_i = \sum_{i \in G_1} a_{ji}x_i + \sum_{i \in G_2} \varepsilon \ \text{sgn}(\alpha_{ii})x_i. \tag{3.5}$$

Suppose x^* is not optimal for (3.2) with the choice of λ_i according to (3.4). Then there exists an $\bar{x} \in S(A, b)$ with

$$\sum_{i=1}^{n} \text{sgn}(\alpha_{ii})\lambda_i \bar{x}_i > \sum_{i=1}^{n} \text{sgn}(\alpha_{ii})\lambda_i x_i^*$$

$$\stackrel{\text{by (3.5)}}{\Longrightarrow} \sum_{i \in G_1} a_{ji}\bar{x}_i + \sum_{i \in G_2} \varepsilon \ \text{sgn}(\alpha_{ii})\bar{x}_i > \sum_{i \in G_1} a_{ji}x_i^* + \sum_{i \in G_2} \varepsilon \ \text{sgn}(\alpha_{ii})x_i^*$$

$$\Rightarrow \sum_{i \in G_1} a_{ji}\bar{x}_i > \sum_{i \in G_1} a_{ji}x_i^* + \sum_{i \in G_2} \varepsilon \ \text{sgn}(\alpha_{ii})[x_i^* - \bar{x}_i]$$

$$\stackrel{\text{by (3.3)}}{\Longrightarrow} \sum_{i \in G_1} a_{ji}\bar{x}_i > b_j - \sum_{i \in G_2} a_{ji}x_i^* + \sum_{i \in G_2} \varepsilon \ \text{sgn}(\alpha_{ii})[x_i^* - \bar{x}_i]. \tag{3.6}$$

By $\bar{x} \in S(A, b)$ we have:

$$\sum_{i \in G_2} a_{ji}\bar{x}_i \quad \leq \quad b_j - \sum_{i \in G_1} a_{ji}\bar{x}_i$$

$$\stackrel{\text{by (3.6)}}{<} \quad \sum_{i \in G_2} a_{ji}x_i^* - \sum_{i \in G_2} \varepsilon \ \text{sgn}(\alpha_{ii})[x_i^* - \bar{x}_i].$$

Thus, for ε sufficiently small we can find an $\underline{i} \in G_2$ such that

$$a_{j\underline{i}}\bar{x}_{\underline{i}} < a_{j\underline{i}}x_{\underline{i}}^* \tag{3.7}$$

$$\stackrel{\text{by } \underline{i} \in G_2}{\Longrightarrow} \alpha_{\underline{i}\underline{i}}\bar{x}_{\underline{i}} > \alpha_{\underline{i}\underline{i}}x_{\underline{i}}^*. \tag{3.8}$$

By (3.3) and (3.7), we have

$$\underbrace{\sum_{\substack{i=1\\i\neq \underline{i}}}^{n} a_{ji}x_i^* + a_{j\underline{i}}\bar{x}_{\underline{i}} = b_j}_{} \underbrace{-a_{j\underline{i}}x_{\underline{i}}^* + a_{j\underline{i}}\bar{x}_{\underline{i}}}_{<0} < b_j.$$

Following this, the solution $(x_{-\underline{i}}^*, \bar{x}_{\underline{i}})$ is feasible with respect to constraint j. If x^* satisfies more than one constraint with equality, the feasibility still holds. In this case due to the same argument, there has to be an $\underline{i}$ in G_2 that satisfies (3.7) for an ε sufficiently small and with respect to all considered constraints. We conclude that by (3.8), x^* is not an equilibrium.

Part (b) Let $\lambda \in \mathbb{R}_+^n : \lambda_i > 0 \; \forall \; i$ with $\alpha_{ii} \neq 0$ and x^* be optimal for (3.2) $\Rightarrow x^*$ equilibrium.

Consider such a solution x^*. Assume that x^* is not an equilibrium. Then

$$\Rightarrow \exists \, i, \exists \, x_i : c_i(x_{-i}^*, x_i) > c_i(x_{-i}^*, x_i^*)$$

$$\Rightarrow \sum_{j=1, j\neq i}^{n} \alpha_{ij}x_j^* + \alpha_{ii}x_i > \sum_{j=1, j\neq i}^{n} \alpha_{ij}x_j^* + \alpha_{ii}x_i^*$$

$$\Rightarrow \alpha_{ii} \neq 0 \; \Rightarrow \; \lambda_i > 0$$

$$\Rightarrow \alpha_{ii}x_i > \alpha_{ii}x_i^* \; \Rightarrow \; \text{sgn}(\alpha_{ii})\lambda_i x_i > \text{sgn}(\alpha_{ii})\lambda_i x_i^*$$

$$\Rightarrow \sum_{j=1, j\neq i}^{n} \text{sgn}(\alpha_{jj})\lambda_j x_j^* + \text{sgn}(\alpha_{ii})\lambda_i x_i$$

$$> \sum_{j=1, j\neq i}^{n} \text{sgn}(\alpha_{jj})\lambda_j x_j^* + \text{sgn}(\alpha_{ii})\lambda_i x_i^*,$$

which contradicts x^* solving (3.2) for all $\lambda \in \mathbb{R}_+^n : \lambda_i > 0 \; \forall \; i$ with $\alpha_{ii} \neq 0$. $\square$

Lemma 3.7. *In a game on a nonempty compact polyhedron $S(A, b)$ with linear payoff functions $c_i(x) = \sum_{j=1}^{n} \alpha_{ij}x_j$, $\alpha_{ij} \in \mathbb{R}$, equilibria do exist.*

Proof. In the trivial case that $\alpha_{ii} = 0 \; \forall \; i = 1, \ldots, n$ holds, each feasible flow is an equilibrium and the claim follows. In all other cases we apply Theorem 3.6. As $S(A, b)$ is compact and the objective function in (3.2) continuous, optimal solutions of (3.2), and thus equilibria, exist. $\square$

The next corollary is given by the fundamental theorem of linear programming theory (see, e.g., [HK00]) and is a direct consequence of Theorem 3.6.

Corollary 3.8. *Consider a game on a nonempty compact polyhedron $S(A, b)$ with linear payoff functions $c_i(x) = \sum_{j=1}^{n} \alpha_{ij}x_j$, $\alpha_{ij} \in \mathbb{R}$, such that there is an i with $\alpha_{ii} \neq 0$. In this game, each equilibrium lies on the boundary of $S(A, b)$, and there is at least one equilibrium that lies in an extreme point of $S(A, b)$.*

3.2.2 Existence and Characterization of Equilibria for Strictly Increasing Payoffs

For games with payoff functions $c_i(x)$ that are strictly increasing in x_i, we show that equilibria exist and present a characterization of the set of equilibria. The first part of the proof of the next theorem uses the idea of the proof concerning Theorem 3.6.

Theorem 3.9. *Consider a game on a compact polyhedron $S(A, b)$, where the payoffs $c_i(x)$ are strictly increasing in x_i. Furthermore, consider the following linear problem,*

$$\max \sum_{i=1}^{n} \lambda_i x_i \quad \text{subject to } x \in S(A, b). \tag{3.9}$$

In this game, the following hold.

(a) If the solution x^* is an equilibrium, then there exist $\lambda_i > 0, i = 1, \ldots, n$, such that x^* solves (3.9).

(b) For all $\lambda_i > 0, i = 1, \ldots, n$, the optimal solution x^* of (3.9) is an equilibrium.

Proof.
Part (a) x^* equilibrium $\Rightarrow \exists \lambda_i > 0, i = 1, \ldots, n$, such that x^* solves (3.9).

As x^* is an equilibrium, it has to lie on the boundary of $S(A, b)$: Suppose not; that is, x^* is an interior point. Rhen there exists a player i and $\delta > 0$ sufficiently small such that $x^\delta = (x_{-i}^*, x_i^* + \delta)$ is in $S(A, b)$. By the strictly increasing payoff functions, $c_i(x^\delta) > c_i(x^*)$ holds which contradicts x^* being an equilibrium.
It follows that there is a constraint j such that it holds:

$$\sum_{i=1}^{n} a_{ji} x_i^* = b_j. \tag{3.10}$$

We fix:

$$\lambda_i = \begin{cases} a_{ji} & \text{if } \text{sgn}(a_{ji}) = 1 \\ \varepsilon & \text{if } \text{sgn}(a_{ji}) \neq 1 \end{cases}, \tag{3.11}$$

with $\varepsilon > 0$, sufficiently small. Furthermore, we define the two sets:

$$G_1 = \{i : \text{sgn}(a_{ji}) = 1\},$$

$$G_2 = \{i : \text{sgn}(a_{ji}) \neq 1\}.$$

Then we get:

$$\sum_{i=1}^{n} \lambda_i x_i = \sum_{i \in G_1} a_{ji} x_i + \sum_{i \in G_2} \varepsilon\, x_i. \qquad (3.12)$$

Suppose x^* is not optimal for (3.9) with the choice of λ_i according to (3.11). Then there exists an $\bar{x} \in S(A,b)$ with

$$\sum_{i=1}^{n} \lambda_i \bar{x}_i > \sum_{i=1}^{n} \lambda_i x_i^*$$

$$\overset{\text{by (3.12)}}{\Longrightarrow} \sum_{i \in G_1} a_{ji} \bar{x}_i + \sum_{i \in G_2} \varepsilon\, \bar{x}_i > \sum_{i \in G_1} a_{ji} x_i^* + \sum_{i \in G_2} \varepsilon\, x_i^*$$

$$\Rightarrow \sum_{i \in G_1} a_{ji} \bar{x}_i > \sum_{i \in G_1} a_{ji} x_i^* + \sum_{i \in G_2} \varepsilon\, [x_i^* - \bar{x}_i]$$

$$\overset{\text{by (3.10)}}{\Longrightarrow} \sum_{i \in G_1} a_{ji} \bar{x}_i > b_j - \sum_{i \in G_2} a_{ji} x_i^* + \sum_{i \in G_2} \varepsilon\, [x_i^* - \bar{x}_i]. \qquad (3.13)$$

By $\bar{x} \in S(A,b)$ we have:

$$\sum_{i \in G_2} a_{ji} \bar{x}_i \quad \le \quad b_j - \sum_{i \in G_1} a_{ji} \bar{x}_i \qquad (3.14)$$

$$\overset{\text{by (3.13)}}{<} \quad \sum_{i \in G_2} a_{ji} x_i^* - \sum_{i \in G_2} \varepsilon\, [x_i^* - \bar{x}_i]. \qquad (3.15)$$

Hence, for ε sufficiently small we can find an $\underline{i} \in G_2$ such that

$$a_{j\underline{i}} \bar{x}_{\underline{i}} < a_{j\underline{i}} x_{\underline{i}}^*$$

$$\Rightarrow a_{j\underline{i}} \neq 0 \overset{\text{by } \underline{i} \in G_2}{\Longrightarrow} a_{j\underline{i}} < 0$$

$$\Rightarrow \bar{x}_{\underline{i}} > x_{\underline{i}}^*.$$

By (3.10) and (3.14), we have

$$\sum_{\substack{i=1 \\ i \neq \underline{i}}}^{n} a_{ji} x_i^* + a_{j\underline{i}} \bar{x}_{\underline{i}} = b_j \underbrace{- a_{j\underline{i}} x_{\underline{i}}^* + a_{j\underline{i}} \bar{x}_{\underline{i}}}_{<0} < b_j.$$

Thus, the solution $(x_{-\underline{i}}^*, \bar{x}_{\underline{i}})$ is feasible with respect to constraint j. If x^* satisfies more than one constraint with equality, the feasibility still holds. In this case due to the same argument, there has to be an $\underline{i}$ in G_2 that satisfies (3.8) for an ε sufficiently small and with respect to all considered constraints. As $c_i(x)$ is strictly increasing in x_i and by (3.15) we have a contradiction to x^* being an equilibrium.

Part (b) Let $\lambda_i > 0, i = 1, \ldots, n$ and x^* be optimal for (3.9) $\Rightarrow$ x^* equilibrium.

Consider a solution x^* of (3.9) for any $\lambda_i > 0, i = 1, \ldots, n$. Assume that x^* is not an equilibrium. Then

$$\Rightarrow \exists\, j, \exists\, x_j : c_j(x^*_{-j}, x_j) > c_j(x^*_{-j}, x^*_j)$$

$$\Rightarrow x_j > x^*_j$$

$$\Rightarrow \sum_{\substack{i=1 \\ i \neq j}}^{n} x^*_i + x_j > \sum_{\substack{i=1 \\ i \neq j}}^{n} x^*_i + x^*_j,$$

which contradicts x^* solving (3.9) for all $\lambda_i > 0, i = 1, \ldots, n$.   □

Lemma 3.10. *In a game on a nonempty compact polyhedron $S(A, b)$ with payoff functions $c_i(x)$ that are strictly increasing in x_i, equilibria do exist.*

Proof. As $S(A, b)$ is compact and the objective function in (3.9) continuous, optimal solutions of (3.9) exist. By Theorem 3.9 the result follows.   □

Corollary 3.11. *In a game on a nonempty compact polyhedron $S(A, b)$ with payoff functions strictly increasing in x_i, each equilibrium lies on the boundary of $S(A, b)$ and there is at least one equilibrium that lies in an extreme point of $S(A, b)$.*

3.2.3 Existence and Characterization of Equilibria for Convex Payoffs

Convex payoffs are not sufficient for the existence of equilibria in games on polyhedra, as the following example shows.

Example 3.12. Consider a game on a compact polyhedron $S(A, b)$ with two players. A and b are given by

$$A = \begin{pmatrix} 1 & 0 \\ 0 & 1 \\ -1 & 0 \\ 0 & -1 \end{pmatrix} \quad b = \begin{pmatrix} 1 \\ 1 \\ 0 \\ 0 \end{pmatrix};$$

that is, $S(A, b)$ is a square. The convex payoffs are given by $c_1(x) = (x_1 + x_2 - 1)^2$ and $c_2(x) = (x_1 - x_2)^2$. Player 1 will try to maximize the distance of the sum of both strategies and the value 1. That means, if player 2 plays a strategy less than 0.5, player 1 will set $x_1 = 0$; if player 2 chooses x_2 to be larger than 0.5, player 1 will set $x_1 = 1$. For $x_2 = 0.5$ player 1 will choose 0 or 1 without preference as both strategies yield the same payoff.

Player 2 will try to maximize the distance between x_1 and x_2; that is, for $x_1 < 0.5$ she will set $x_2 = 1$ and for $x_1 > 0.5$ she will choose $x_2 = 0$. Also for $x = 0.5$ she chooses x_2 equal one or zero.

In this game no equilibrium exists. If $|x_1 - x_2| < 0.5$, player 2 will have incentive to change her strategy (possibly also player 1 wants to change strategy, but not necessarily). For $|x_1 - x_2| > 0.5$ at least player 1 has incentive for changing the strategy. If we have $|x_1 - x_2| = 0.5$, either both players want to change strategy (if $x_1, x_2 \in (0, 1)$), or the player that plays 0.5 has incentive for changing the strategy.

Lemma 3.13. *Consider a game on a polyhedron such that there is a player k with payoff $c_k(x)$ strictly convex in x_k. In such a game, all equilibria lie on the boundary of the polyhedron, if they exist.*

Proof. Assume there is an equilibrium x^* which is an interior point of $S(A, b)$. Then there is an $\varepsilon > 0$, sufficiently small, such that $(x^*_{-k}, x^*_k \pm \varepsilon) \in S(A, b)$. Either $d = v_k$ or $d = -v_k$ (with v_k being the kth unit vector) is an improvement direction. Thus, we can find a solution (x^*_{-k}, x_k) such that $c_k(x^*_{-k}, x_k) > c_k(x^*_{-k}, x^*_k)$ which contradicts the assumption.   □

3.2.4 Nondominated Equilibria

The definition of dominance in games on polyhedra keeps to the classical definition of dominance in multicriteria literature (see, e.g., [Ehr05]). Let $u \in \mathbb{R}^k$ and $v \in \mathbb{R}^k$. We write $u \gneqq v$ if $u_i \geq v_i \; \forall \; i = 1, \dots, k$ holds and if there is a j such that $u_j > v_j$ is true. We recall Definition 2.68 (page 50) from path player games, in the notation of games on polyhedra.

Definition 3.14. *A solution $\hat{x}$ is called* dominated *if there exists a dominating solution; that is, a solution x such that*

$$c(x) \gneqq c(\hat{x}).$$

Otherwise, $\hat{x}$ is called nondominated.

Dominance with respect to a particular game on a polyhedron, namely the path player game is described in Section 2.3.

The set of nondominated solutions in a game is denoted by $ND(\Gamma)$. This set is interesting, because it contains those solutions that are reasonable for the players. Solutions outside $ND(\Gamma)$ are not preferable as we can find at least one solution in $ND(\Gamma)$ such that no player receives a smaller payoff and at least one player is able to increase his payoff. It is now of special interest to compare the set of equilibria, which we denote with $NE(\Gamma)$, with the set $ND(\Gamma)$. Already for linear payoffs and in games with two players, the four variants: $ND(\Gamma) = NE(\Gamma)$, $ND(\Gamma) \subset NE(\Gamma)$, $ND(\Gamma) \supset NE(\Gamma)$, and $ND(\Gamma) \cap NE(\Gamma) = \emptyset$ may occur. We demonstrate this in the following examples.

Example 3.15 ($ND(\Gamma) = NE(\Gamma)$).
We consider a two-player game on a polyhedron with two players and the linear payoff functions $c_1(x) = x_1$ and $c_2(x) = x_2$. Figure 3.5 illustrates polyhedra where $ND(\Gamma) = NE(\Gamma)$ holds. The three examples show that the sets lie on one or several faces of the polyhedron.

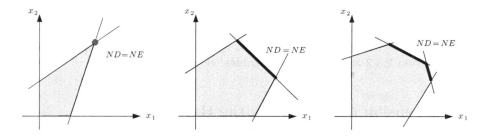

Fig. 3.5. $ND(\Gamma) = NE(\Gamma)$.

Example 3.16 ($ND(\Gamma) \subset NE(\Gamma)$).
For a game on a polyhedron with two players and linear payoff functions $c_1(x) = C$, C constant and $c_2(x) = x_2$ we obtain $ND(\Gamma) \subset NE(\Gamma)$ for a polyhedron as illustrated in Figure 3.6. Here $ND(\Gamma)$ consist of a single solution whereas $NE(\Gamma)$ is described by two facets.

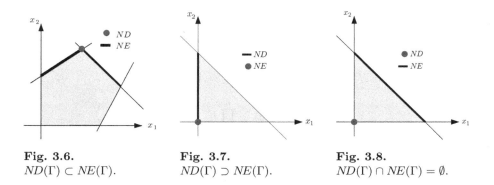

Fig. 3.6.
$ND(\Gamma) \subset NE(\Gamma)$.

Fig. 3.7.
$ND(\Gamma) \supset NE(\Gamma)$.

Fig. 3.8.
$ND(\Gamma) \cap NE(\Gamma) = \emptyset$.

Example 3.17 ($ND(\Gamma) \supset NE(\Gamma)$).
For a game on a polyhedron with two players and linear payoff functions $c_1(x) = -x_1 + 2x_2$ and $c_2(x) = -x_2$ we obtain $ND(\Gamma) \supset NE(\Gamma)$ for the

polyhedron illustrated in Figure 3.7. Here, $NE(\Gamma) = \{(0,0)\}$ and $ND(\Gamma) = \{x \in S : x_1 = 0\}$.

Example 3.18 ($ND(\Gamma) \cap NE(\Gamma) = \emptyset$).
For a game on a polyhedron with two players and linear payoff functions $c_1(x) = 1/2x_1 - x_2$ and $c_2(x) = -x_1 + 1/2x_2$ we obtain $ND(\Gamma) \neq NE(\Gamma)$ for the polyhedron illustrated in Figure 3.8.

As path player games (with no security limit) are instances of games on polyhedra, Example 2.75 on page 56 shows that $ND(\Gamma) \nsubseteq NE(\Gamma) \wedge ND(\Gamma) \nsupseteq NE(\Gamma) \wedge ND(\Gamma) \cap NE(\Gamma) \neq \emptyset$ may also hold (Note that the other examples from Section 2.3.2 are also valid for games on polyhedra).

3.3 Extension to a Game on the Hypercuboid

In this section, we describe a way to transform games on polyhedra to a new type of game, namely to games on hypercuboids. The intention of this approach is to get rid of the nonstatic strategy sets that are for games on polyhedra related by constraints. Instead, we create fixed strategy sets for all players, which are consequently easier to analyze. In the following, we also call the game on a polyhedron the *original game* and the game on the hypercuboid the *extended game*. After formally defining these games, we investigate the relation between equilibria in these games.

For our purpose, we consider the smallest hypercuboid $H(A, b)$ that contains the polyhedron $S(A, b)$.

Definition 3.19. *Given a game on a polyhedron $S(A,b)$, we define the* strategy set *for the corresponding extended game on the hypercuboid $H(A,b)$ by*

$$h_i = \bigcup_{x_{-i} \in \mathbb{R}^{n-1}} S_i(x_{-i}).$$

The hypercuboid $H(A,b)$ *corresponding to $S(A,b)$ is given by*

$$H(A,b) = \prod_{i=1}^{n} h_i.$$

The set h_i contains all x_i such that a feasible strategy vector (x_i, x_{-i}) can be found. Thus, h_i may alternatively be described as

$$h_i = \left\{ x_i : \exists\, x_{-i} \in \mathbb{R}^{n-1} \text{ such that } (x_{-i}, x_i) \in S(A,b) \right\}.$$

Note that $H(A,b)$ is unbounded if and only if $S(A,b)$ is unbounded.

In most cases, the strategy set of a game on the hypercuboid will be larger than that of the corresponding original game on the polyhedron. Thus, the definition of payoffs has to be extended. The payoff for a game on a hypercuboid is defined such that it represents the original payoff of the game on the polyhedron for all solutions within $S(A, b)$. For infeasible solutions outside the polyhedron, however, a penalty is introduced. In the following, we assume that the cost $c_i : H(A, b) \to \mathbb{R}$ is given on the complete hypercuboid.

In fact, the assumption to have $c = (c_i)_{i=1,\dots,n}$ given over the complete $H(A, b)$ is realistic. In many restricted problems, such as questions from production planning or in traffic optimization, we can expect to have a cost or payoff function given for a general domain, for instance, the nonnegative cone $\mathbb{R}^n_+$, whereas later the set of feasible solutions is described by a set of linear or nonlinear constraints. We consider just linear constraints and obtain a polyhedric feasible set $S(A, b)$.

Definition 3.20. *Given a game on the polyhedron $S(A, b)$ with payoff function $c : H(A, b) \to \mathbb{R}^n$. The* payoff *for the corresponding extended game on the hypercuboid $H(A, b)$ is for $x \in H(A, b)$ given by*

$$c_i^H(x) = \begin{cases} c_i(x) & \text{if} \quad x \in S(A, b) \\ -M + c_i(x) & \text{else} \end{cases},$$

where M is a large number, at least

$$M > \max_{\substack{i=1,\dots,n \\ x \in S(A,b)}} c_i(x).$$

If we fix the strategies of the competitors x_{-i}, we can rewrite the payoff of any player i such that it depends only on x_i. We denote this one-dimensional payoff by $\tilde{c}_i^H(x_i)$.

$$\tilde{c}_i^H(x_i) = \begin{cases} c_i(x_{-i}, x_i) & \text{if} \quad x_i \in S_i(x_{-i}) \\ -M + c_i(x_{-i}, x_i) & \text{else} \end{cases}.$$

Definition 3.21. *A solution $x^* \in H(A, b)$ is an* equilibrium *in a game on a hypercuboid if and only if for all players $i = 1, \dots, n$ it holds that*

$$c_i^H(x^*_{-i}, x_i^*) \geq c_i^H(x^*_{-i}, x_i) \quad \forall \, x_i \in h_i.$$

Note that the game on a hypercuboid is not generalized anymore. Furthermore, a game on a hypercuboid is a game on a polyhedron itself, although a degenerated one.

The following lemmas describe relations between the original equilibria in the game on polyhedra and the corresponding extended game on hypercuboids.

Lemma 3.22. *A solution x^* is an equilibrium in a game on a hypercuboid $H(A, b)$, if it is an equilibrium in the corresponding game on the polyhedron $S(A, b)$.*

Proof. As x^* is an equilibrium in the game on the polyhedron $S(A, b)$ we get

$$\Rightarrow \ c_i(x^*_{-i}, x^*_i) \geq c_i(x^*_{-i}, x_i) \quad \forall \, i, \, \forall \, x_i \in S_i(x_{-i})$$

$$\Rightarrow c_i^H(x^*_{-i}, x^*_i) \geq c_i^H(x^*_{-i}, x_i) \quad \forall \, i, \, \forall \, x_i \in S_i(x_{-i}) \qquad (3.16)$$

$$\Rightarrow c_i^H(x^*_{-i}, x^*_i) \geq c_i^H(x^*_{-i}, x_i) \quad \forall \, i, \, \forall \, x_i \in h_i, \qquad (3.17)$$

and thus, x^* is an equilibrium in the game on $H(A, b)$.

Inequality (3.16) is true due to $x^* \in S(A, b)$, and (3.17) holds because $c_i^H(x) = -M + c_i(x)$ for all solutions x within $H(A, b)$ but outside $S(A, b)$.

$\square$

The reverse conclusion of Lemma 3.22 does not hold in general, as the following example illustrates.

Example 3.23. Consider a game on a polyhedron with three players and

$$A = \begin{pmatrix} 1 & 1 & 1 \\ -1 & 0 & 0 \\ 0 & -1 & 0 \\ 0 & 0 & -1 \end{pmatrix} \quad b = \begin{pmatrix} 1 \\ 0 \\ 0 \\ 0 \end{pmatrix}.$$

The polyhedron $S(A, b)$ is presented in Figure 3.9. The corresponding hypercuboid $H(A, b)$ is given by the strategy sets $h_i = [0, 1], i = 1, 2, 3$.

Given the payoff functions $c_i(x) = x_i$, $i = 1, 2, 3$, the solution $x^* = (1, 1, 1)$ is an equilibrium in the game on $H(A, b)$ due to

$$c_i(x^*_{-i}, x_i) = -M + x_1 \leq -M + 1 = c_i(x^*_{-i}, x^*_i) \ \forall \ x_i \in h_i, \ i = 1, 2, 3.$$

However, x^* is no equilibrium in the game on $S(A, b)$, because x^* is not feasible.

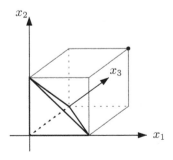

Fig. 3.9. Infeasible equilibrium in a game on a hypercuboid.

Nevertheless, equilibria in a game on a hypercuboid correspond to equilibria in a game on the polyhedron if they are feasible.

Lemma 3.24. *A solution x^* is an equilibrium in a game on $S(A, b)$, if it is an equilibrium in the corresponding game on $H(A, b)$ and $x^* \in S(A, b)$ holds.*

Proof. As $x^* \in S(A, b)$, it holds that $c_i^H(x^*) = c_i(x^*)$. From x^* being an equilibrium in the game on $H(A, b)$ we obtain

$$c_i^H(x^*_{-i}, x^*_i) \geq c_i^H(x^*_{-i}, x_i) \; \forall \, i, \; \forall \, x_i \in h_i$$
$$\Rightarrow c_i^H(x^*_{-i}, x^*_i) \geq c_i^H(x^*_{-i}, x_i) \; \forall \, i, \; \forall \, x_i \in S_i(x_{-i})$$
$$\Rightarrow c_i(x^*_{-i}, x^*_i) \geq c_i(x^*_{-i}, x_i) \quad \forall \, i, \; \forall \, x_i \in S_i(x_{-i}),$$

thus, x^* is an equilibrium in the game on $S(A, b)$. $\qquad\Box$

Theorem 3.25. *A feasible solution x^* is an equilibrium in a game on $S(A, b)$ if and only if x^* is an equilibrium in the game on the corresponding hypercuboid $H(A, b)$.*

Proof. The claim follows from Lemma 3.22 and Lemma 3.24. $\qquad\Box$

Generally, we may have more equilibria in the extended game than in the original game, because we may have infeasible equilibria in the former. It is useful to determine games where no infeasible equilibria in the game on the hypercuboid exist; that is, the sets of equilibria are equal in the original and in the extended game. Then, an equilibrium for the original generalized game can be found by solving the extended game, a game with independent strategy sets.

A game where this is true is the two-player version of the game on a polyhedron.

Lemma 3.26. *Consider a game on a polyhedron $S(A, b)$ with two players. A solution x^* is an equilibrium in that game if and only if it is an equilibrium in the corresponding game on the hypercuboid $H(A, b)$.*

Proof.
Part (a) x^* equilibrium in $S(A, b)$ $\Rightarrow$ x^* equilibrium in $H(A, b)$.
See Lemma 3.22.
Part (b) x^* equilibrium in $H(A, b)$ $\Rightarrow$ x^* equilibrium in $S(A, b)$.
Consider the set $S_1(x_{-1}) = S_1(x_2) = \{x_1 : A(x_1, x_2)^T \leq b\}$.

$$\forall \, x_2 \in h_2 \qquad \exists \, x_1 : x_2 \in S_2(x_1) \qquad \text{[by Definition 3.19]}$$
$$\Rightarrow \forall \, x_2 \in h_2 \qquad \exists \, x_1 : (x_1, x_2) \in S(A, b)$$
$$\Rightarrow \forall \, x_2 \in h_2 \text{ holds:} \qquad S_1(x_2) \neq \emptyset.$$

Now assume that $x^* \notin S(A, b)$. It follows that $c_i^H(x^*) = -M \; \forall \, i = 1, 2$. But as $S_1(x_2^*)$ is nonempty, it can be concluded that we can find an $x_1 \in S_1(x_2^*)$

such that $c_1^H(x_1^*, x_2^*) = -M < c_1(x_1, x_2^*) = c_1^H(x_1, x_2^*)$; that is, x^* is not an equilibrium. This contradicts the assumption and we conclude that $x^* \in S(A, b)$. By Lemma 3.24, x^* is an equilibrium for the game on $S(A, b)$.   □

3.4 Potentials for Games on Polyhedra

3.4.1 Potentials for Games on Polyhedra in the Extension to the Hypercuboid

Before we introduce potential functions for games on the hypercuboid, we recall the main definitions and results from Section 2.4.1, the introduction to potential functions for path player games, in the notation of games on polyhedra.

Definition 3.27. *A function* $\Pi : H(A, b) \to \mathbb{R}$ *is an* exact potential function *for a game* Γ *if for every* $i = 1, \ldots, n$, *for every* x_{-i}, *and for every* $x_i^1, x_i^2 \in h_i$ *it holds:*

$$c_i^H(x_{-i}, x_i^1) - c_i^H(x_{-i}, x_i^2) = \Pi(x_{-i}, x_i^1) - \Pi(x_{-i}, x_i^2). \qquad (3.18)$$

By a result of [MS96], a solution x^* that maximizes $\Pi(x)$ is an equilibrium for a game on the hypercuboid $H(A, b)$; compare Lemma 2.81 on page 61. That means, if we find a potential function for a game on $H(A, b)$ and if the maximum of this potential exists, we can conclude the existence of equilibria in this game. Moreover, the potential function will help us to determine equilibria.

We now investigate games on hypercuboids that have exact potential functions. As we show later, the maximizers of the exact potential functions presented in this section have to lie in $S(A, b)$. Thus, for bounded $S(A, b)$ we can even guarantee the existence of feasible equilibria in these cases.

Definition 3.28. *A* strategy sequence $\varphi = (x^0, x^1, \ldots, x^k, \ldots)$ *(or simply sequence) in a game is given as an ordered sequence of solutions* $x^k \in H(A, b)$ *that satisfies for every* $k \geq 1$: *there is a unique player* $i(k)$ *such that for*

$$x^{k-1} = \left(x_{-i(k)}^{k-1}, x_{i(k)}^{k-1} \right) \quad and \quad x^k = \left(x_{-i(k)}^k, x_{i(k)}^k \right)$$

we have

$$x_{-i(k)}^{k-1} = x_{-i(k)}^k \quad and \quad x_{i(k)}^{k-1} \neq x_{i(k)}^k.$$

We call $i(k)$ *the* active player *and the* movement from x^{k-1} to x^k *the* kth step *in* φ.

A sequence is called an improvement sequence *if for every* $k \geq 1$ *it holds that*

$$c_{i(k)}^H \left(x^k \right) > c_{i(k)}^H \left(x^{k-1} \right). \qquad (3.19)$$

A sequence is feasible *if* $x^k \in S(A, b) \ \forall \ k \geq 0$. *For a sequence* φ, *we call* x^0 *its* initial solution *and, if* φ *is finite,* x^N *its* terminal solution. *Furthermore, we say that a finite* φ *is* connecting x^0 *and* x^N. *The* length *of a finite sequence* $\varphi = (x^0, x^1, \ldots, x^N)$ *is given by* $l(\varphi) = N$.

The cost of a finite sequence $\varphi = (x^0, x^1, \ldots, x^N)$, $x^k \in H(A, b)$ is analogous to Definition 2.84 for games on polyhedra; see page 62. However, for games on hypercuboids, we have to distinguish between the cost of a sequence with respect to payoff c^H and the cost of a sequence with respect to payoff c.

Definition 3.29. *Consider a finite sequence* $\varphi = (x^0, x^1, \ldots, x^N)$, $x^k \in H(A, b)$. *The* cost *of* φ *with respect to payoff* c^H *is given as*

$$I^H(\varphi) = \sum_{k=1}^{N} \left[c_{i(k)}^H(x^k) - c_{i(k)}^H(x^{k-1}) \right]. \tag{3.20}$$

Moreover, the cost *of* φ *with respect to payoff* c *is given by*

$$I^c(\varphi) = \sum_{k=1}^{N} \left[c_{i(k)}\left(x^k\right) - c_{i(k)}\left(x^{k-1}\right) \right]. \tag{3.21}$$

Note that both expressions, (3.20) and (3.21), are valid for $x^k \in H(A, b)$, although in (3.21) we do not consider the correct payoff c_i^H for a game on the hypercuboid. Instead, we consider the much simpler payoff c_i.

Definition 3.30. *A sequence is* closed *if* $x^0 = x^N$ *holds and a closed sequence is* simple *if* $x^\ell \neq x^k$ *is true for all* $\ell \neq k$ *and* $0 \leq \ell, k \leq N - 1$.

With the following lemma, we recall Lemma 2.87 on page 62 by Monderer and Shapley [MS96].

Lemma 3.31. *Consider a game* Γ *on a hypercuboid* $H(A, b)$. *The following statements are equivalent.*

Γ *is an exact potential game.* $\hspace{2cm}$ (3.22)

$I^H(\varphi) = 0$ *for every finite closed sequence* φ *in* $H(A, b)$. $\hspace{0.5cm}$ (3.23)

$I^H(\varphi) = 0$ *for every finite simple closed sequence* φ *in* $H(A, b)$. (3.24)

$I^H(\varphi) = 0$ *for every simple closed sequence* φ *in* $H(A, b)$

$\hspace{2cm}$ *of length 4 .* $\hspace{2cm}$ (3.25)

An exact potential of Γ is given by fixing a solution $\bar{x} \in H(A, b)$ and defining $\Pi(x) = I^H(\varphi) \ \forall \ x \in H(A, b)$ where φ is a sequence connecting $\bar{x}$ and x. Note that $\Pi(x)$ is well defined, as $I^H(\varphi_1) = I^H(\varphi_2)$ holds for φ_1 and φ_2 having the same initial and terminal solution.

For the classical definition of exact potential functions as presented in [MS96], independent strategy sets are required. Thus, we cannot apply it directly to generalized games. An adapted definition of potential functions that works for generalized games is called *restricted potential function* and was proposed in Section 2.4.2. We apply this definition to games on polyhedra in Section 3.4.2. The extension to games on hypercuboids, however, enables static strategy sets and thus allows us to use the classical definition of exact potentials in this section.

The first observation is unfortunately a negative one. Games on hypercuboids are not exact potential games in general. The next example illustrates this fact.

Example 3.32. Consider the game on the polyhedron $S(A,b)$, presented in Example 3.5 on page 90, and the closed sequence

$$\varphi^1 = ((0,0);(0,1);(0.9,1);(0.9,0);(0,0)).$$

This sequence is feasible, such that $I^H(\varphi^1)$ and $I^c(\varphi^1)$ are equal in this case. The cost of this sequence is given by

$$
\begin{aligned}
I^H(\varphi^1) &= I^c(\varphi^1) \\
&= (1-0) + (-0.01 - (-1)) + (0.89 - 0.01) + (0 - (-0.89)) \\
&= 3.76 \neq 0.
\end{aligned}
$$

By Lemma 3.31, a potential function cannot exist in the corresponding extended game on $H(A,b)$. Note that as φ^1 is feasible, we cannot expect to have potential functions in the original game on the polyhedron (potential functions for games on polyhedra are defined in Section 3.4.2). This relation is described more formally in Theorem 3.39, page 111 .

In Example 3.5 we have observed that the game on $S(A,b)$ has no equilibrium. But, also for games with existing equilibria, the existence of potential functions is not necessarily given: Consider the game on the polyhedron $S(\bar{A},\bar{b})$ presented in the same example. Here, equilibria do exist and, for example, the closed and feasible sequence

$$\varphi^2 = ((0,0);(0,1);(1,1);(1,0);(0,0))$$

with cost

$$I^H(\varphi^2) = I^c(\varphi^2) = (1-0) + (-0 - (-1)) + (1-0) + (0 - (-1)) = 4 \neq 0$$

indicates that the game on $H(\bar{A},\bar{b})$ does not provide a potential function.

To prove the existence of potential functions for special instances of games on hypercuboids, one can use Lemma 3.31 by showing that all simple closed sequences in $H(A,b)$ of length four have cost zero with respect to c^H.

The following lemma uses this idea and shows that for games on hypercuboids it is even sufficient to check that the cost of these sequences is zero with respect to the much simpler cost function c.

Lemma 3.33. *A game on a hypercuboid $H(A, b)$ with payoff function c_i^H is an exact potential game if and only if the game on the hypercuboid $H(A, b)$ with changed payoff c_i (leaving off penalty $-M$) is an exact potential game.*

Proof. Consider a game on a hypercuboid $H(A, b)$ and two players K and L that create a simple closed sequence φ in $H(A, b)$ of length four. The players start with the strategies x_K, x_L, change once to $\bar{x}_K$ and $\bar{x}_L$, and then back. The strategies of the remaining players $x_{-\{KL\}}$ stay constant, such that the sequence is given by

$$\varphi = ((x_{-\{KL\}}, x_K, x_L), (x_{-\{KL\}}, \bar{x}_K, x_L), (x_{-\{KL\}}, \bar{x}_K, \bar{x}_L),$$
$$(x_{-\{KL\}}, x_K, \bar{x}_L), (x_{-\{KL\}}, x_K, x_L)).$$

To simplify the notation, we denote the solutions in the sequence as

$$\varphi = ((KL), (\bar{K}L), (\bar{K}\bar{L}), (K\bar{L}), (KL)).$$

Consider the cost of φ with respect to c, given as

$$I^c(\varphi) = c_K(\bar{K}L) - c_K(KL)$$
$$+c_L(\bar{K}\bar{L}) - c_L(\bar{K}L) \tag{3.26}$$
$$+c_K(K\bar{L}) - c_K(\bar{K}\bar{L})$$
$$+c_L(KL) - c_L(K\bar{L}).$$

Consider next the cost of φ with respect to c^H. Let x be a solution in φ, $\hat{x}$ the successor in the sequence, and i the active player in the step from x to $\hat{x}$. It holds:

$$c_i^H(\hat{x}) - c_i^H(x) = c_i(\hat{x}) - c_i(x) + \xi(x, \hat{x})M,$$

where

$$\xi(x, \hat{x}) = \begin{cases} -1 & \text{if } x \notin S(A, b), \hat{x} \in S(A, b) \\ 1 & \text{if } x \in S(A, b), \hat{x} \notin S(A, b) \\ 0 & \text{otherwise} \end{cases}.$$

Thus, the cost of sequence φ with respect to c^H is given by

$$I^H(\varphi) = c_K(\bar{K}L) - c_K(KL) + \xi((KL), (\bar{K}L)) M$$
$$+c_L(\bar{K}\bar{L}) - c_L(\bar{K}L) + \xi((\bar{K}L), (\bar{K}\bar{L})) M$$
$$+c_K(K\bar{L}) - c_K(\bar{K}\bar{L}) + \xi((\bar{K}\bar{L}), (K\bar{L})) M$$
$$+c_L(KL) - c_L(K\bar{L}) + \xi((K\bar{L}), (KL)) M$$
$$= I^c(\varphi)$$
$$+M \left(\xi((KL), (\bar{K}L)) + \xi((\bar{K}L), (\bar{K}\bar{L})) \right.$$
$$+\xi((\bar{K}\bar{L}), (K\bar{L})) + \xi((K\bar{L}), (KL)) \big).$$

As φ is closed, the numbers of steps from feasible to infeasible solutions equals the number of steps from infeasible to feasible solutions and hence

$$\xi\left((KL),(\bar{K}L)\right) + \xi\left((\bar{K}L),(\bar{K}\bar{L})\right) + \xi\left((\bar{K}\bar{L}),(K\bar{L})\right) + \xi\left((K\bar{L}),(KL)\right) = 0.$$

Hence we have

$$I^H(\varphi) = 0 \quad \Leftrightarrow \quad I^c(\varphi) = 0.$$

Thus, by Lemma 3.31 the claim follows. $\square$

Using the previous lemma, we are now able to prove the existence of potential functions for special instances of payoff functions.

Lemma 3.34. *Consider a game on a hypercuboid $H(A,b)$ with linear payoffs $c_i(x) = \sum_{j=1}^n \alpha_{ij} x_j$, $\alpha_{ij} \in \mathbb{R}$ on $H(A,b)$. This game is an exact potential game.*

Proof. By Lemma 3.33 it remains to show that each closed simple sequence φ in $H(A,b)$ of length four has cost zero with respect to c. By the linearity of the payoffs we get for all $(x^1_{-i}, x_i), (x^1_{-i}, \bar{x}_i), (x^2_{-i}, x_i),$ and $(x^2_{-i}, \bar{x}_i) \in H(A,b)$:

$$c_i(x^1_{-i}, x_i) - c_i(x^1_{-i}, \bar{x}_i) = c_i(x^2_{-i}, x_i) - c_i(x^2_{-i}, \bar{x}_i).$$

Thus,

$$c_K(x_{-\{KL\}}, \bar{x}_K, x_L) - c_K(x_{-\{KL\}}, x_K, x_L)$$
$$= -\left(c_K(x_{-\{KL\}}, x_K, \bar{x}_L) - c_K(x_{-\{KL\}}, \bar{x}_K, \bar{x}_L)\right)$$

and

$$c_L(x_{-\{KL\}}, \bar{x}_K, \bar{x}_L) - c_L(x_{-\{KL\}}, \bar{x}_K, x_L)$$
$$= -\left(c_L(x_{-\{KL\}}, x_K, x_L) - c_L(x_{-\{KL\}}, x_K, \bar{x}_L)\right)$$

holds. Since $I^c(\varphi)$ is given as presented in (3.26), we conclude that $I^c(\varphi) = 0$. $\square$

Another class of games on hypercuboids providing potential functions is given by games with the *path player game(PPG)* property.

Definition 3.35. *We denote with $\mathbb{P}(n)$ the power set of the set of players $\{1, \ldots, n\}$. A game on a polyhedron $S(A,b)$ has the* path player game (PPG) *property if for all $m \in \mathbb{P}(n)$ there exists a function $z_m(x_m)$ with $x_m = (x_j)_{j\in m}$ such that for all $i = 1, \ldots, n$ it holds*

$$c_i(x) = \sum_{m\in\mathbb{P}(n):i\in m} z_m(x_m). \tag{3.27}$$

A game on a hypercuboid $H(A,b)$ has the PPG property, if $S(A,b)$ has.

To satisfy the PPG property, a game needs a special payoff structure. We have to analyze the payoff function $c_i(x)$ and group the components that are dependent on the decisions of the same set of players m, denoted by $z_m(x_m)$. If it is possible to decompose the payoffs of all players $c_i(x)$, $i = 1, \ldots, n$ into a set of functions $z_m(x_m)$ such that these functions are equal for all players i, the game has the PPG property. In other words, all components in a payoff function that are influenced by a set of competitors m contribute to the payoffs of these competitors with the same additive component.

The definition is motivated by the class of path player games (see Chapter 2), which is an instance of games on polyhedra. In Section 2.4, the existence of potential functions for path player games is provided. In fact, each path player game is a game on a polyhedron with the PPG property, but the reverse is not true. The definition of the PPG property is more general, as it allows arbitrary components $z_m(x_m)$ whereas for a path player game we need to have $z_m(x_m) = z_m \left(\sum_{i \in m} x_i \right)$.

Example 3.36. Consider a game on a polyhedron $S(A, b)$ with three players. The payoffs are given by

$$c_1(x) = x_1^2 + 2(x_1 + x_2) + x_1 x_3 - (x_2 x_3^2 + x_1)^2,$$
$$c_2(x) = x_2 + 2(x_1 + x_2) - (x_2 x_3^2 + x_1)^2,$$
$$c_3(x) = 3x_3 + x_1 x_3 - (x_2 x_3^2 + x_1)^2.$$

This game has the PPG property, because the payoffs can be decomposed into

$$z_{\{1\}}(x) = x_1^2, \ z_{\{2\}}(x) = x_2, \ z_{\{3\}}(x) = 3x_3,$$
$$z_{\{1,2\}}(x) = 2(x_1 + x_2), \ z_{\{1,3\}}(x) = x_1 x_3,$$
$$z_{\{1,2,3\}}(x) = -(x_2 x_3^2 + x_1)^2.$$

Nevertheless, this game is not a path player game, as the components $z_m(\cdot)$ depend on arbitrary expressions and not only on sums of the strategies.

The decomposition of cost functions is in general not as simple as in Example 3.36 and moreover, it need not be unique. In fact it is a nontrivial problem to find a decomposition into components $z_m(\cdot)$.

Lemma 3.37. *A game on a hypercuboid with the PPG property is an exact potential game.*

Proof. By Lemma 3.33 it suffices to show that each simple closed sequence φ^S in $H(A, b)$ of length four has cost zero with respect to c. Consider two players K and L that create such a sequence φ. The cost $I^c(\varphi)$ is given as in (3.26). Inserting (3.27) we get the following equation.

We denote with $x_{m,-i} = (x_\ell)_{\ell \in m \setminus \{i\}}$ the vector that contains the strategies associated with the elements of m, except the element i. We define $x_{m,-\{ij\}} = (x_\ell)_{\ell \in m \setminus \{i,j\}}$ analogously.

$$I^c(\varphi) =$$

$$\left(\sum_{m\in\mathbb{P}(n):K\in m, L\notin m} z_m(x_{m,-K}, \bar{x}_K) + \sum_{m\in\mathbb{P}(n):K,L\in m} z_m(x_{m,-\{KL\}}, \bar{x}_K, x_L) \right)$$

$$- \left(\sum_{m\in\mathbb{P}(n):K\in m, L\notin m} z_m(x_{m,-K}, x_K) + \sum_{m\in\mathbb{P}(n):K,L\in m} z_m(x_{m,-\{KL\}}, x_K, x_L) \right)$$

$$+ \left(\sum_{m\in\mathbb{P}(n):L\in m, K\notin m} z_m(x_{m,-L}, \bar{x}_L) + \sum_{m\in\mathbb{P}(n):K,L\in m} z_m(x_{m,-\{KL\}}, \bar{x}_K, \bar{x}_L) \right)$$

$$- \left(\sum_{m\in\mathbb{P}(n):L\in m, K\notin m} z_m(x_{m,-L}, x_L) + \sum_{m\in\mathbb{P}(n):K,L\in m} z_m(x_{m,-\{KL\}}, \bar{x}_K, x_L) \right)$$

$$+ \left(\sum_{m\in\mathbb{P}(n):K\in m, L\notin m} z_m(x_{m,-K}, x_K) + \sum_{m\in\mathbb{P}(n):K,L\in m} z_m(x_{m,-\{KL\}}, x_K, \bar{x}_L) \right)$$

$$- \left(\sum_{m\in\mathbb{P}(n):K\in m, L\notin m} z_m(x_{m,-K}, \bar{x}_K) + \sum_{m\in\mathbb{P}(n):K,L\in k} z_m(x_{m,-\{KL\}}, \bar{x}_K, \bar{x}_L) \right)$$

$$+ \left(\sum_{m\in\mathbb{P}(n):L\in m, K\notin m} z_m(x_{m,-L}, x_L) + \sum_{m\in\mathbb{P}(n):K,L\in m} z_m(x_{m,-\{KL\}}, x_K, x_L) \right)$$

$$- \left(\sum_{m\in\mathbb{P}(n):L\in m, K\notin m} z_m(x_{m,-L}, \bar{x}_L) + \sum_{m\in\mathbb{P}(n):K,L\in m} z_m(x_{m,-\{KL\}}, x_K, \bar{x}_L) \right)$$

$$= \quad 0.$$

$\square$

Note that this theorem is an extension of Theorem 2.96 (page 68).

Thus far we have presented two classes of games on hypercuboids that provide exact potentials. Note that Example 2.88 on page 63 is not a contradiction to these results, although it has linear payoffs and the PPG property. To make sure that potentials exist, we have to consider path player games with extended benefit functions, as introduced in Section 2.4.4.

In both classes of games presented a potential $\Pi(x)$ is given by the cost of any sequence connecting an arbitrarily chosen, but fixed, $\hat{x} \in H(A, b)$ with x. The existence of an exact potential in a game on $H(A, b)$ is a basis of proving the existence of equilibria in games on hypercuboids. Nevertheless, for this result, we need in addition the definition of equilibria in the original game on $S(A, b)$. Therefore we postpone the existence statement to Section 3.4.2; see Lemma 3.42.

3.4.2 Restricted Potentials for Games on Polyhedra in the Original Game

The game on a polyhedron in its original version is a generalized Nash equilibrium game. GNE games are not considered in the classical definition of potential games. In Section 2.4.2 we introduced the exact restricted potential function. In the following, we transfer this definition to GNE games. For games on polyhedra, we show the relation between exact potential games on hypercuboids and exact restricted potential games on polyhedra. We take advantage of this relation by presenting instances of games on polyhedra that are exact restricted potential games. Furthermore, we prove the existence of equilibria in games on compact polyhedra, and thus also in games on compact hypercuboids for (restricted) potential games.

Definition 3.38. *A function* $\Pi : S(A,b) \to \mathbb{R}$ *is an* exact restricted potential function *for a GNE game* Γ *if for every* $i = 1, \ldots, n$, *for all feasible* x_{-i}, *and for all* x_i^1, $x_i^2 \in S_i(x_{-i})$ *it holds:*

$$c_i(x_{-i}, x_i^1) - c_i(x_{-i}, x_i^2) = \Pi(x_{-i}, x_i^1) - \Pi(x_{-i}, x_i^2). \tag{3.28}$$

A GNE game Γ *is called an* exact restricted potential game *if it admits an exact restricted potential.*

Theorem 3.39. *Consider a polyhedron* $S(A,b)$. *Let* Γ^S *be the original game on* $S(A,b)$ *and* Γ^H *be the extended game on* $H(A,b)$. *The game* Γ^S *is an exact restricted potential game if* Γ^H *is an exact potential game.*

Proof. As Γ^H is an exact potential game, we find by Definition 3.27, page 104, an exact potential function Π such that

$$c_i^H(x_{-i}, x_i^1) - c_i^H(x_{-i}, x_i^2) = \Pi(x_{-i}, x_i^1) - \Pi(x_{-i}, x_i^2) \tag{3.29}$$

for all $i = 1, \ldots, n$ and for all $(x_{-i}, x_i^1), (x_{-i}, x_i^2) \in H(A,b)$. In particular (3.29) holds for all $(x_{-i}, x_i^1), (x_{-i}, x_i^2) \in S(A,b)$. As we have $c_i^H(x) = c_i(x)$ for all $x \in S(A,b)$ and for all $i = 1, \ldots, n$, it follows by Definition 3.38 that Π is also an exact restricted potential function for the orginial game on $S(A,b)$. □

It is not possible to prove the reverse implication of the theorem above due to the following reason. Given a game on a polyhedron $S(A,b)$ together with a payoff $c : S(A,b) \to \mathbb{R}^n$ that provides an exact restricted potential function, it is possible to define $c : H(A,b) \to \mathbb{R}^n$ such that the potential property is destroyed outside $S(A,b)$.

Games on polyhedra are in general not exact restricted potential games, just as games on hypercuboids are not necessarily exact potential games. This observation is provided by Example 3.32 on page 106 as we find there a closed simple sequence φ^1 of length four lying entirely in $S(A,b)$ with cost

$I^H(\varphi^1) = I^c(\varphi^1) \neq 0$. But, if an exact restricted potential function existed for the game on the polyhedron $S(A, b)$, then for all closed finite sequences φ of length four within $S(A, b)$ it would hold $I^c(\varphi) = 0$. This is true due to the following equations, which hold by the definition of an exact restricted potential function.

$$I^c(\varphi) = \sum_{k=1}^{4} \left[c_{i(k)}(f^k) - c_{i(k)}(f^{k-1}) \right]$$

$$= \sum_{k=1}^{4} \left[\Pi(f^k) - \Pi(f^{k-1}) \right]$$

$$= \Pi(f^4) - \Pi(f^0) = 0.$$

Lemma 3.40. *Consider a game Γ^{lin} on a polyhedron $S(A, b)$ containing linear payoffs $c_i(x) = \sum_{j=1}^{n} \alpha_{ij} x_j$, $\alpha_{ij} \in \mathbb{R}$. Furthermore, consider a game Γ^{PPG} that has the PPG property.*
The games Γ^{lin} and Γ^{PPG} are exact restricted potential games.

Proof. The claim follows from Theorem 3.39, Lemma 3.34, and Lemma 3.37.
□

Lemma 3.41. *Consider a game on a compact polyhedron $S(A, b)$, which provides an exact restricted potential function. In this game, equilibria exist.*

Proof. As $S(A, b)$ is compact, the intervals $S_i(x_{-i})$ are also compact for all $i = 1, \ldots, n$. By Definition 3.38 and as we have continuous payoff functions $c_i(x)$, the exact restricted potential function $\Pi(x)$ is continuous. It follows that the maximum of $\Pi(x), x \in S(A, b)$ exists. The maximizer of the potential function is an equilibrium; see a result of [MS96]. Thus, the claim follows.   □

Note that Monderer and Shapley present in [MS96] a similar proof for infinite potential games with independent strategy sets; that is, for nongeneralized games.

Lemma 3.42. *Consider a game Γ^H on a hypercuboid $H(A, b)$, obtained from a game Γ^S on a compact polyhedron $S(A, b)$. Assume that Γ^H provides an exact potential function. Then, feasible equilibria exist in Γ^H.*

Proof. As Γ^H is an exact potential game, it follows that Γ^S is an exact restricted potential game; see Theorem 3.39. By Lemma 3.41, equilibria exist in Γ^S, which are feasible equilibria in Γ^H by Lemma 3.22.   □

Corollary 3.43. *Consider a game Γ^{lin} on a polyhedron $S(A, b)$ with linear payoffs $c_i(x) = \sum_{j=1}^{n} \alpha_{ij} x_j$, $\alpha_{ij} \in \mathbb{R}$. Furthermore, consider a game Γ^{PPG} that has the PPG property.*
In the original games Γ^{lin} and Γ^{PPG}, and in their extensions to hypercuboids, equilibria exist.

For linear payoffs this statement has already been provided in Lemma 3.7, together with a complete characterization of the set of equilibria in Theorem 3.6. For path player games, existence is given for all instances of the game; see the three different approaches for existence proofs in Theorem 2.31 (page 24), Theorem 2.103 (page 73), and Theorem 2.118 (page 84).

Regarding the computation of equilibria in games on polyhedra with the PPG property, we present the following result.

Theorem 3.44. *For a game on a polyhedron $S(A,b)$ which satisfies the PPG property, an equilibrium is given by an optimal solution of the following problem.*

$$\max \sum_{m \in \mathbb{P}(n)} [z_m(x_m) - z_m(0)] \quad \text{subject to } x \in S(A,b).$$

Proof. In the following we show that a potential function for games on polyhedra with the PPG property and for $x \in S(A,b)$ is given by

$$\Pi(x) = \sum_{m \in \mathbb{P}(n)} [z_m(x_m) - z_m(0)].$$

Then, the maximizer of $\Pi(x) : x \in S(A,b)$ is an equilibrium and the claim follows.

Consider a game on a polyhedron with n players, satisfying the PPG property. Let

$$\varphi = \left(\begin{pmatrix} 0 \\ \vdots \\ 0 \end{pmatrix}, \begin{pmatrix} x_1 \\ 0 \\ \vdots \\ 0 \end{pmatrix}, \dots, \begin{pmatrix} x_1 \\ \vdots \\ x_{n-1} \\ 0 \end{pmatrix}, \begin{pmatrix} x_1 \\ \vdots \\ x_n \end{pmatrix} \right)$$

be the sequence, connecting $\bar{x} = \mathbf{0}_n$ and x. Denote with

$$I_k^c(\varphi) = c_{i(k)}(x^k) - c_{i(k)}(x^{k-1})$$

the "cost" of the kth step. By the definition of the PPG property (see (3.27)),

$$I_k^c(\varphi) = \sum_{m \in \mathbb{P}(n):i(k)\in m} z_m\left(x_m^k\right) - \sum_{m \in \mathbb{P}(n):i(k)\in m} z_m\left(x_m^{k-1}\right)$$

$$= z_{\{i(k)\}}\left(x_{\{i(k)\}}^k\right) + \sum_{\substack{m \in \mathbb{P}(n):i(k)\in m \\ m \neq \{i(k)\}}} z_m\left(x_m^k\right)$$

$$- \left(z_{\{i(k)\}}\left(x_{\{i(k)\}}^{k-1}\right) + \sum_{\substack{m \in \mathbb{P}(n):i(k)\in m \\ m \neq \{i(k)\}}} z_m\left(x_m^{k-1}\right) \right).$$

As $x_{\{i(k)\}}^k = x_{\{i(k)\}}$ and $x_{\{i(k)\}}^{k-1} = 0$ hold, we get:

$$I^c(\varphi) = \sum_{k=1}^{n} \left[z_{\{i(k)\}} \left(x_{\{i(k)\}} \right) - z_{\{i(k)\}}(0) \right]$$

$$+ \sum_{k=1}^{n} \left[\sum_{\substack{m \in \mathbb{P}(n): i(k) \in m \\ m \neq \{i(k)\}}} \left(z_m \left(x_{m,\{i(k)\}}, x^k_{m,-\{i(k)\}} \right) - z_m \left(0, x^{k-1}_{m,-\{i(k)\}} \right) \right) \right]}_{(\#)}.$$

We reorder $(\#)$ by exchanging the summations:

$$(\#) = \sum_{\substack{m \in \mathbb{P}(n): i(k) \in m \\ m \neq \{i(k)\}}} \left[\sum_{i(k) \in m} \left(z_m \left(x_{m,\{i(k)\}}, x^k_{m,-\{i(k)\}} \right) - z_m \left(0, x^{k-1}_{m,-\{i(k)\}} \right) \right) \right].$$

As $x^k_q = x_q$ holds for $i(k) \geq q$, and $x^k_q = 0$ for $i(k) < q$ we get:

$$(\#) = \sum_{\substack{m \in \mathbb{P}(n): i(k) \in m \\ m \neq \{i(k)\}}} \left[z_m(x_m) - z_m(0) \right].$$

Summarizing, we have:

$$I^c(\varphi) = \sum_{m \in \mathbb{P}(n)} \left[z_m(x_m) - z_m(0) \right].$$

For an exact restricted potential game Γ an exact restricted potential function is derived by fixing a feasible solution $\bar{x} \in S(A, b)$ and defining $\Pi(x) = I^c(\varphi) \ \forall \ x \in S$ where φ is a feasible sequence connecting $\bar{x}$ and x. Thus,

$$\Pi(x) = \sum_{m \in \mathbb{P}(n)} \left[z_m(x_m) - z_m(0) \right]. \qquad \square$$

3.4.3 Computation of Equilibria by Improvement Sequences

Improvement sequences, such as best-reply improvement sequences (see Definition 2.110, page 78), yield, if they are maximal and finite, equilibria as terminal solutions. For path player games, we have been able to present instances that satisfy the finite best-reply property (FBRP); that is, each best-reply improvement sequence is finite for these games. If this is true, algorithms based on best-reply improvement sequences can be constructed for computation of equilibria. For example, this is used in Algorithm 2 in Section 4.5.1.

Unfortunately, for games on polyhedra, we do not get such a result, inasmuch as FBRP may not be satisfied even for games with linear, strictly increasing cost functions (that are hence exact restricted potential games). We present such an example next.

Example 3.45. Consider the game on the polyhedron $S(A, b)$ with

$$A = \begin{pmatrix} -1 & 2 \\ 2 & -1 \\ -1 & 0 \\ 0 & -1 \end{pmatrix} \qquad b = \begin{pmatrix} 2 \\ 2 \\ 0 \\ 0 \end{pmatrix}.$$

The costs are given by $c_1(x_1) = x_1$ and $c_2(x_2) = x_2$. For a given $\bar{x}_2$, the first player will choose $x_1 = 1 + 0.5\bar{x}_2$ as the best reply, and for a given $\bar{x}_1$, the second player chooses $x_2 = 1 + 0.5\bar{x}_1$. The only fixed point of this mapping is given at $(2, 2)$. All best-reply improvement sequences starting at $x \in S(A, b) \setminus \{(2, 2)\}$ are infinite and hence FBRP does not hold. See the illustration in Figure 3.10.

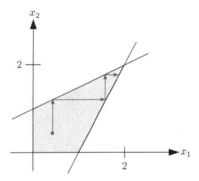

Fig. 3.10. Best-reply sequence is infinite.

Another approach to improvement sequences is that of ε-improvement sequences (Definition 2.114, page 82), which yield approximate equilibria (Definition 2.115). A game has an approximate finite improvement property (AFIP) if for all $\varepsilon > 0$ each ε-improvement sequence is finite. For games on polyhedra that have exact restricted potential functions, we can prove the finiteness.

Lemma 3.46. *A game on a polyhedron that is an exact restricted potential game satisfies AFIP.*

Proof. As $c_i(x)$ are continuous for all $i = 1, \ldots, n$ and given over a closed interval it follows that the payoff is bounded. By Definition 3.38, the exact restricted potential function is bounded, too. As a feasible ε-improvement sequence improves $\Pi(f)$ by at least ε in each step, each feasible ε-improvement sequence has to be finite.   □

Thus, using ε-improvement sequences, ε-equilibria with a given precision can be obtained for games on polyhedra with exact restricted potential functions.

4

The Line Planning Game: An Application

Given is a public transportation network $G = (V, E)$ consisting of stations $v \in V$ and direct connections between stations $e \in E$. Given, furthermore, a set of possible lines, find a line plan; that is, decide which of the lines should be established and with what frequency.

4.1 Introduction

The *line planning game (LPG)* models the process of establishing lines in a public transport system, for example, a railway or a bus network. In line planning, a *public transportation network (PTN)* is modeled by vertices for each stop (as with a train station) and edges for each direct connection between stops (as with tracks between stations). A *line* is given as a path in the PTN, and the *frequency* indicates how often the bus or train goes within a certain time horizon. Usually, the installed lines have to satisfy a certain amount of demand (e.g., provide enough resources to carry the customers who want to travel in the PTN). On the other hand, the amount of traffic is limited (e.g., by safety regulations). Problems of this kind can be solved with respect to the cost of operating lines, but customer-oriented objectives, such as short traveling times or changes between lines are also considered.

In [BKZ96], the focus is on maximizing the number of passengers with direct connections, whereas in [SS05, Sch05, BGP04a, BGP04b] traveling times are minimized.

In our approach, we consider the line planning problem from the game-theoretic point of view. The lines act as players who choose as a strategy their frequency and keep this information private in the game. Their objective is to minimize an expected delay, which is dependent on the frequencies of all lines in the network, and is represented by the payoff of this game.

Usually, a traffic network consists of several origin–destination(OD) pairs. In order to keep the notation clear, we consider in our model just a single OD

S. Schwarze, *Path Player Games*, DOI 10.1007/978-0-387-77928-7_4,
© Springer Science+Business Media, LLC 2009

pair. In Section 4.5.2 we briefly discuss how the model can be extended to multiple OD pairs.

In this chapter, we show that the line planning game is a generalization of the path player game introduced in Chapter 2. On the other hand, the LPG is an instance of games on polyhedra, which are defined and discussed in Chapter 3. We use this fact to transfer results from games on polyhedra to line planning games. In particular, we show that potential functions exist for the generalized version of line planning games. Furthermore, the existence of equilibria is given and we develop three methods for computation of equilibria (see Theorems 4.17–4.21). Finally, we demonstrate one method within a numerical example using data from the German railway system.

The material presented in this section is not only an illustration of the practical use of the methods developed in the earlier sections, but also a starting point for further research in the field of transport optimization. It is planned to investigate open questions in this field in the framework of the European project, "Algorithms for robust and online railway optimization: Improving the validity and reliability of large scale systems (ARRIVAL)" [arr].

4.2 The Line Planning Game Model

We consider a network $G = (V, E)$ with vertices $v \in V$ and edges $e \in E$, where V and E are nonempty and finite. A *line* P in G is given by a finite path of edges $e \in E$: $P = (e_1, \ldots, e_k)$. We denote the *line pool* $\mathcal{P}$ as a set of lines P in G from the single *origin* s to the single *destination* t. We use in the line planning chapter the notations "origin" and "destination" instead of "source" and "sink", because this is more common in transport optimization. Note that in contrast to the path player game, $\mathcal{P}$ does not necessarily contain all lines from s to t. Assigned to each edge $e \in E$ is a cost function $c_e(\cdot)$ that describes the expected average delay on that edge. Naturally, this function depends on the load on e. We assume the cost functions to be continuous and nonnegative for nonnegative load; that is, $c_e(x) \geq 0$ for $x \geq 0$. We make no further assumption on the cost functions, although in line planning the costs are usually nondecreasing.

The *frequency* on a line P is denoted by f_P. The frequencies in the complete network are represented by the *frequency vector*, given by $f : \mathcal{P} \to \mathbb{R}_+$. Consequently, the frequency on an edge $e \in E$ is given by the sum of the frequencies on lines that contain e,

$$f_e = \sum_{P:e\in P} f_P. \tag{4.1}$$

The cost on a line P is given by the sum of costs on the edges belonging to that line,

$$c_P(f) = \sum_{e \in P} c_e(f_e).$$

As is common in the literature about line planning, we have to consider two types of bounds regarding the frequencies. First, a *minimal frequency* from s to t has to be covered to meet the demand of the customers. This minimal frequency is given by $f^{\min} \geq 0$ and we require the sum of all frequencies on lines from s to t to cover the demand; that is,

$$\sum_{P \in \mathcal{P}} f_P \geq f^{\min}. \tag{4.2}$$

If the condition is not satisfied; that is, if there is not enough frequency in the network, all lines receive a payoff M, with M being a large number working as a penalty.

The second bound is the real-valued *maximal frequency* $0 \leq f_e^{\max} < \infty$ that is assigned to each edge $e \in E$; that is, it has to hold

$$f_e \leq f_e^{\max} \ \forall \ e \in E. \tag{4.3}$$

The maximal frequency establishes a capacity constraint on the edges and in line planning problems is usually given by security issues. Too much traffic on a railway could cause a collapsing system, therefore the constraint is a measure of precaution. If the frequency f_e exceeds $f_e^{\max}$; that is, if $f_e > f_e^{\max}$, all lines that contain e will receive a payoff of $N < M$. We allow N to be any real value smaller than M, nevertheless in the line planning problem, we choose N being a large number to create a punishment for exceeding flow.

Definition 4.1. *A line planning game* Γ *is given by the tuple*

$$\Gamma = (G, \mathcal{P}, f^{\min}, f^{\max}, c, N, M).$$

By using (4.1) we rewrite the constraint (4.3), $\sum_{P:e \in P} f_P \leq f_e^{\max}$, and obtain the following definition.

Definition 4.2. *A frequency vector f is called* feasible *if the bounds $f^{\min}$ and $f_e^{\max}, e \in E$, are satisfied. The set of feasible frequency vectors is given by*

$$\mathbb{F}^{\mathrm{LPG}} = \left\{ f \in \mathbb{R}_+^{|\mathcal{P}|} : \sum_{P \in \mathcal{P}} f_P \geq f^{\min} \ \wedge \ \sum_{P:e \in P} f_P \leq f_e^{\max} \ \forall \ e \in E \right\}.$$

The above definition differs from the set of feasible flows $\mathbb{F}$ in path player games, presented in (2.3) on page 24, as no bounds on edges have been considered there.

Definition 4.3. *The* payoff function *of a line P is for $f \geq \mathbf{0}_{|\mathcal{P}|}$ given by*

$$b_P(f) = \begin{cases} c_P(f) & \text{if } \sum_{P_k \in \mathcal{P}} f_{P_k} \geq f^{\min} \wedge \forall \ e \in P : f_e \leq f_e^{\max} \\ N & \text{if } \sum_{P_k \in \mathcal{P}} f_{P_k} \geq f^{\min} \wedge \exists \ e \in P : f_e > f_e^{\max} \\ M & \text{if } \sum_{P_k \in \mathcal{P}} f_{P_k} < f^{\min} \end{cases}.$$

Each line tries to minimize its own payoff which depends on the frequencies of all lines f. Note that M is a punishment that concerns all players, whereas N concerns only those players owning the corresponding edge.

To illustrate the payoff function of line P, we fix for a given frequency vector f the frequencies f_{P_k} with $P_k \neq P$. We obtain the frequency vector f_{-P}, by deleting the Pth component in the frequency vector f. The payoff function depending only on f_P is illustrated in Figure 4.1.

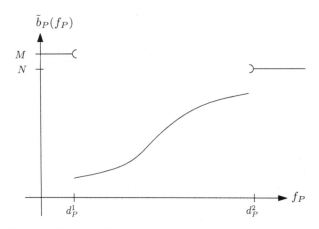

Fig. 4.1. Payoff of line P for a fixed frequency vector f_{-P}.

The payoff consists of three continuous intervals. The left part is described by the payment M in the case of not satisfying the minimal frequency $f^{\min}$. The middle part is given by the sum of costs on the edges belonging to P. If we have nondecreasing cost functions c_e, which are usual in line planning problems, the sum $c_P(f)$ of nondecreasing functions is also nondecreasing. The right part is given by the payoff N which is paid if the maximal frequency is violated on an edge belonging to the line P. The values that mark the boundaries of the intervals, denoted by d_P^1 and d_P^2, depend on the bounds $f^{\min}$ and $f_e^{\max}, e \in E$ and on the frequencies f_{-P} of the opponents. The exact definition is given by the following.

Definition 4.4. *In a line planning game, the* lower decision limit *is given by*

$$d_P^1(f_{-P}) = f^{\min} - \sum_{P_k \in \mathcal{P} \backslash \{P\}} f_{P_k}. \tag{4.4}$$

The upper decision limit *is denoted with*

$$d_P^2(f_{-P}) = \min_{e \in P} \{f_e^{\max} - f_{e,-P}\}, \tag{4.5}$$

with $f_{e,-P} = f_e - f_P$.

If no confusion regarding the chosen f_{-P} arises, we denote the lower and the upper decision limits by d_P^1 and d_P^2, respectively.

The lower decision limit is given by the minimal frequency less than the frequencies that are already provided by the competitors. Thus, given the flow of the competitors, a player P should at least set a frequency of d_P^1 or higher. For the upper decision limit, $f_{e,-P}$ is given by the flow on e that is produced by the remaining players in $\mathcal{P}$ apart from P. For each edge in P we consider the difference of the maximal frequency and the frequency that is already used by the lines of the competitors. Because the maximal frequency shall not be violated on any edge, the minimal value of these differences provides the limit d_P^2. In fact, the existence of two bounds in $b_P(f)$ that depend on the flow f_{-P} is one of the main differences from the path player game. There we had for each path P the decision limit $d_P(f_{-P})$ dependent on the flow of the other players as the upper bound and ω_P as the fixed lower bound.

As we have $N \gg 0$ and $M \gg 0$ in the LPG, each line should choose a frequency in $[d_P^1, d_P^2] \cap \mathbb{R}_+$, in order to minimize its payoff. By choosing from $[d_P^1, d_P^2]$, the constraints (4.2) and (4.3) are satisfied. Moreover, the obtained f is feasible if we consider nonnegative frequencies in addition. It may happen that $[d_P^1, d_P^2] \cap \mathbb{R}_+$ is empty for a player P in a specific game situation, even if $\mathbb{F}^{\mathrm{LPG}}$ is nonempty. This may be caused by the other players, who have already violated the constraints (4.2) and (4.3) such that P is not able to create a feasible flow.

An equilibrium in the line planning game is given by the following.

Definition 4.5. *In a line planning game, a frequency vector f^* is an equilibrium if and only if for all lines $P \in \mathcal{P}$ and for all $f_P \geq 0$ it holds that*

$$b_P(f_{-P}^*, f_P^*) \leq b_P(f_{-P}^*, f_P).$$

This definition of equilibria considers that in a line planning game, the players want to minimize the payoff, whereas in path player games, the payoff is maximized. Equilibria in line planning games may be feasible or infeasible, which can be observed in the following example. As we are interested in implementable solutions, we analyze feasible frequencies.

Example 4.6. We consider a line planning game with a line pool containing two lines. Let f_1 and f_2 be the frequencies of these lines. The minimal frequency $f^{\mathrm{min}} = 1$ has to be covered from s to t. The game network consists of three edges, as illustrated in Figure 4.2. The maximal frequencies of the edges are given by $f_{e_1}^{\mathrm{max}} = f_{e_2}^{\mathrm{max}} = 2$ and $f_{e_3}^{\mathrm{max}} = 3$. Furthermore, the following costs are assigned to the edges: $c_{e_1}(x) = x$, $c_{e_2}(x) = 2x$, and $c_{e_3}(x) = x^2$. Thus, we obtain payoffs: $c_1(f) = f_1 + (f_1 + f_2)^2$ for the first player and $c_2(f) = 2f_2 + (f_1 + f_2)^2$ for the second player. See Figure 4.3 for an illustration of the set of feasible frequencies $\mathbb{F}^{\mathrm{LPG}}$.

This line planning game provides multiple equilibria. Feasible equilibria are, for example, $f^1 = (1, 0)$ and $f^2 = (0, 1)$, with payoffs $b(f^1) = (2, 1)$ and $b(f^2) = (1, 3)$. There are also infeasible equilibria, for example, $f^3 = (4, 4)$, where no player is able to receive a smaller payoff than N.

The frequency vector $f^4 = (3, 3)$ is no equilibrium, although no player is able to reach the set of feasible frequencies within one step. It is a property of line planning games (which is not shared with path player games) that outside the feasible region each player does not necessarily get punished. Here, for example, player 1 could change his frequency to zero. The resulting frequency vector $\bar{f}^4 = (0, 3)$ is still infeasible, but player 1 is able to improve his payoff from $b_1(f^4) = N$ to $b_1(\bar{f}^4) = 9$.

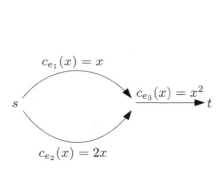

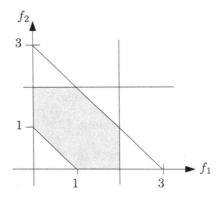

Fig. 4.2.
Game network of Example 4.6.

Fig. 4.3.
Set of feasible frequencies $\mathbb{F}^{\mathrm{LPG}}$.

The above example illustrates that in line planning games, there may be areas of infeasible frequency vectors, where some players violate constraints and others do not. See Figure 4.4, the illustration of the two-player game of Example 4.6, and consider the four infeasible regions A, B, C, and D. For frequency vectors that lie in region A, both players get punished with payoff N, whereas in region B, both receive the payoff M. In region C, only player 1 gets punished with payoff N, whereas player 2 receives a payoff $c_2(f)$, as player 2 satisfies the maximal frequencies $f_e^{\max}$ on her edges e_2 and e_3, but player 1 violates $f_{e_1}^{\max} = 2$. In region D the reverse situation takes place: player 2 gets punished and player 1 does not. Situations such as in regions C and D happen because the players are not sharing the same set of constraints. A systematic investigation of such areas for standard networks $G(n)$ (see Section 2.1.5) is a topic of future research.

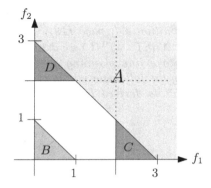

Fig. 4.4. Infeasible regions A, B, C, and D.

The line planning game is a basic model for analyzing line planning problems with game-theoretical means. Other aspects of line planning could be incorporated in extended versions of the game:

–   Consider setup costs for installing lines (or fixed costs for operating a line) and/or consider a maximal number of lines to install.
–   Consider length (e.g., travel time) of a line, as this is a natural measure of attractiveness of a line.
–   Allow passengers to change the lines.

A drawback of the basic model is that frequencies f_P are real numbers, whereas in practice, only $f_P \in \mathbb{N}_0$ make sense. We consider an extension to an integer LPG in Section 4.5.1.

4.3 The Path Player Game as an Instance of the Line Planning Game

In this section, we show that the line planning game is an extension of the path player game; that is, each path player game is an instance of a line planning game. For this purpose, we define the following transformation of any path player game to a line planning game. Afterwards, we show in Theorem 4.8 that the equilibria in the original and the transformed game coincide.

The idea of the transformation is the following. To each path in the path player game an artificial edge e_0^P is added as the first edge. An artificial source $\bar{s}$ is added to the set of vertices, and the artificial edges are inserted as parallel edges from $\bar{s}$ to s. These new edges carry the security limits ω_P of paths $P \in \mathcal{P}$ as maximal frequency $f_{e_0^P}^{\max}$ and thus transfer the bound on a path to a bound on an edge. Each artificial edge is owned exclusively by line P and we set $c_{e_0^P} = 0$ and $f_{e_0^P}^{\max} = \omega_P$ for all P.

Definition 4.7. *Consider a PPG, given by* $\hat{\Gamma} = (\hat{G}, \hat{\mathcal{P}}, \hat{r}, \hat{\omega}, \hat{c}, \hat{\kappa}, \hat{M})$ *(we denote PPG analogous to Definition 4.1, page 119), with corresponding flow* $\hat{f}$ *and benefit functions* $\hat{b}_P(f)$. *With* $\Gamma = T^1(\hat{\Gamma})$ *we denote the* T^1-*transformation of the path player game into a line planning game* $\Gamma = (G, \mathcal{P}, f^{\min}, f^{\max}, c, N, M)$ *that is described by the following relations.*

The network $G = (V, E)$ *is derived from* $\hat{G} = (\hat{V}, \hat{E})$ *by the following modifications.*

$$V = \hat{V} \cup \{\bar{s}\},$$

$$E = \hat{E} \cup \{e_0^{P_1}, \dots, e_0^{P_n}\} \quad \text{with } e_0^{P_k} = (\bar{s}, s) \ \forall P_k \in \mathcal{P}, \text{ where } s$$
$$\text{is the single source in } \hat{\Gamma} \text{ and } \bar{s} \text{ the single source in } \Gamma.$$

For each path $\hat{P} = (e_1^{\hat{P}}, \dots, e_{m_{\hat{P}}}^{\hat{P}}) \in \hat{\mathcal{P}}$, *we construct a line* $P = (e_0^P, e_1^P, \dots, e_{m_P}^P)$ *with* $e_i^P = e_i^{\hat{P}}$, $\forall \ i = 1, \dots, m_P$. *Finally, we obtain the line pool* $\mathcal{P}$ *as the set of lines* P. *The following relations complete the transformation.*

$$f = -\hat{f},$$
$$c_e(f_e) = \begin{cases} -\hat{c}_e(-f_e) = -\hat{c}_e(\hat{f}_e) & \text{if } e \in \hat{G} \\ 0 & \text{otherwise} \end{cases},$$
$$f^{\min} = -\hat{r},$$
$$f_e^{\max} = \begin{cases} \infty & \text{if } e \in \hat{G} \\ -\hat{\omega}_P & \text{if } e = e_0^P \end{cases}, \tag{4.6}$$
$$N = -\hat{\kappa}_P$$
$$M = \hat{M}.$$

Figure 4.5 illustrates a path player game network $\hat{\Gamma}$ where two paths P and $\tilde{P}$ are highlighted. In Figure 4.6 the network of the corresponding line planning game is presented. This network includes the artificial sink $\bar{s}$ and the artificial edges e_0^P.

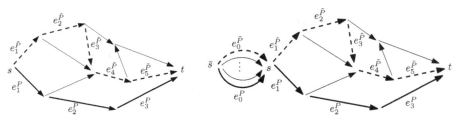

Fig. 4.5. PPG network $\hat{\Gamma}$.   **Fig. 4.6.** LPG network Γ.

Before we prove the equivalence of the games Γ and $\hat{\Gamma}$, we have to point out one difficulty of the T^1-transformation. In path player games we have cost functions $\hat{c}_e(\hat{f}_e)$ that are defined only for $f_e \geq 0$ and which are required to be nonnegative. Hence we obtain costs c_e for the line planning game, which are given for $f_e \leq 0$ and which are nonpositive. Moreover, we have nonpositive $f^{\min}$ and $f_e^{\max}$, due to ω_P and r being nonnegative. However, according to the definition of a line planning game, nonnegative variables and parameters are required. We cope with that problem by a second transformation that shifts the frequencies f_P as well as the cost functions $c_P(\cdot)$. We discuss this second transformation in Lemma A.1 in Appendix A.

Theorem 4.8. *Consider the games* $\hat{\Gamma}$ *and* $\Gamma = T^1(\hat{\Gamma})$ *and the frequencies* $\hat{f}$ *in* $\hat{G}$ *and* $f = -\hat{f}$ *in* G. *It holds that* $b_P(f) = -\hat{b}_P(\hat{f})$ *for all* $P \in \mathcal{P}$.

Proof. For the proof we check the equality of $b_P(f)$ and $-\hat{b}_P(\hat{f})$ for the three components of the benefit and payoff function.

Part (a)

$$b_P(f) = M \Leftrightarrow \sum_{P \in \mathcal{P}} f_P < f^{\min} \Leftrightarrow \sum_{P \in \mathcal{P}} -\hat{f}_P < f^{\min}$$

$$\Leftrightarrow \sum_{P \in \mathcal{P}} \hat{f}_P > -f^{\min} \Leftrightarrow \sum_{P \in \mathcal{P}} \hat{f}_P > \hat{r}$$

$$\Leftrightarrow \hat{b}_P(\hat{f}) = -M.$$

Part (b)

$$b_P(f) = c_P(f) \Leftrightarrow \underbrace{\sum_{P \in \mathcal{P}} f_P \geq f^{\min}}_{(*)} \wedge \underbrace{\forall e \in P : f_e \leq f_e^{\max}}_{(**)}$$

$$\Leftrightarrow \sum_{P \in \mathcal{P}} \hat{f}_P \leq \hat{r} \wedge \hat{f}_P \geq \hat{\omega}_P$$

$$\Leftrightarrow \hat{b}_P(\hat{f}) = \hat{c}_P(\hat{f}) = \sum_{e \in P} \hat{c}_e(\hat{f}_e) = \sum_{e \in P} -c_e(-\hat{f}_e) = -c_P(f).$$

$$(*) \quad \sum_{P \in \mathcal{P}} f_P \geq f^{\min} \Leftrightarrow \sum_{P \in \mathcal{P}} -\hat{f}_P \geq f^{\min} \Leftrightarrow \sum_{P \in \mathcal{P}} \hat{f}_P \leq -f^{\min}$$

$$\Leftrightarrow \sum_{P \in \mathcal{P}} \hat{f}_P \leq \hat{r}.$$

$$(**) \; \forall e \in P : f_e \leq f_e^{\max} \Leftrightarrow f_{e_0^P} \leq f_{e_0^P}^{\max} \quad [\text{by definition of } f_e^{\max}]$$

$$\Leftrightarrow f_{e_0^P} \leq -\hat{\omega}_P$$

$$\Leftrightarrow f_P \leq -\hat{\omega}_P$$

[holds as e_0^P is owned exclusively by P]

$$\Leftrightarrow -\hat{f}_P \leq -\hat{\omega}_P \Leftrightarrow \hat{f}_P \geq \hat{\omega}_P.$$

Part (c)

$$b_P(f) = N \Leftrightarrow \underbrace{\sum_{P \in \mathcal{P}} f_P \geq f^{\min}}_{(*)} \wedge \underbrace{\exists\, e \in P : f_e > f_e^{\max}}_{(***)}$$

$$\Leftrightarrow \sum_{P \in \mathcal{P}} \hat{f}_P \leq \hat{r} \wedge \hat{f}_P < \hat{\omega}_P$$

$$\Leftrightarrow \hat{b}_P(\hat{f}) = \hat{\kappa}_P = -N.$$

$(*)$ see Part (b)

$(***)\ \exists\, e \in P : f_e > f_e^{\max} \Leftrightarrow f_{e_0^P} > f_{e_0^P}^{\max}$

[" $\Rightarrow$ " holds as $f_e^{\max} = \infty \ \forall e \in P \setminus \{e_0^P\}$]

$\Leftrightarrow f_P > f_{e_0^P}^{\max}$ [as $f_P = f_{e_0^P}$ holds]

$\Leftrightarrow -\hat{f}_P > -\hat{\omega}_P \Leftrightarrow \hat{f}_P < \hat{\omega}_P.$ $\square$

Corollary 4.9. *The flow $\hat{f}$ solves $\max \hat{b}_P(\hat{f})$ for all $P \in \mathcal{P}$ if and only if $f = -\hat{f}$ solves $\min b_P(f)$ for all $P \in \mathcal{P}$. That is, $\hat{f}$ is an equilibrium in $\hat{\Gamma}$ if and only if $f = -\hat{f}$ is one in Γ.*

The presented results yield, in particular, two conclusions. First, the LPG is a generalization of the path player game. Hence we can apply any result we obtain for line planning games also to path player games. Second, if we have a line planning game Γ that is obtained by the transformation $T^1(\hat{\Gamma})$ for a path player game $\hat{\Gamma}$, we can transfer results from path player games to line planning games. In the following, we present two other instances of line planning games that can be obtained from a path player game.

Maximal frequencies are infinite

Consider a line planning game where for each line $P \in \mathcal{P}$ it holds: all edges $e_P \in P$ provide maximal frequencies that are given by $f_{e_P}^{\max} = \infty$, apart from one edge $\bar{e}_P \in P$ which may have an arbitrary value $f_{\bar{e}_P}^{\max} \geq 0$. Edge $\bar{e}_P$ has to be owned exclusively by P; that is, $\bar{e}_P \notin P_k \ \forall\, P_k \in \mathcal{P} \setminus \{P\}$. This game corresponds to a path player game, and is connected to it by the same relations (4.6) as used in the T^1-transformation.

Line-disjoint network with equal maximal frequencies

Consider an LPG $\Gamma = (G, \mathcal{P}, f^{\min}, f^{\max}, c, N, M)$ where G is a line-disjoint network; that is, a network where each edge belongs to exactly one line. Assume that for all lines P the following holds: $\forall e_1, e_2 \in P : f_{e_1}^{\max} = f_{e_2}^{\max}$. Due to the line-disjoint network it holds that $\forall e \in P : f_e = f_P$.

From the LPG, the corresponding PPG $\hat{\Gamma} = (\hat{G}, \hat{\mathcal{P}}, \hat{r}, \hat{c}, \hat{\omega}, \hat{\kappa}, \hat{M})$ is obtained by setting the parameters as follows.

$$
\begin{aligned}
\hat{G} &= G, \\
\hat{\mathcal{P}} &= \mathcal{P}, \\
\hat{f} &= -f, \\
\hat{c}_e(\hat{f}_e) &= -c_e(-\hat{f}_e), \\
\hat{\omega}_P &= -f_e^{\max}, \quad \text{with } e \in P, \\
\hat{r} &= -f^{\min}, \\
\hat{\kappa}_P &= -N \\
\hat{M} &= M.
\end{aligned}
$$

Theorem 4.10. *Consider the frequencies $\hat{f}$ in $\hat{\Gamma}$ and $f = -\hat{f}$ in Γ. It holds that $b_P(f) = -\hat{b}_P(\hat{f})$ for all $P \in \mathcal{P}$. Thus, $\hat{f}$ is an equilibrium in $\hat{\Gamma}$ if and only if $f = -\hat{f}$ is an equilibrium in Γ.*

We prove this claim analogously to Theorem 4.8, with two adjustments in Part (b) and Part (c).

Part (b)

$$
\begin{aligned}
(**) \; \forall e \in P : f_e \leq f_e^{\max} &\Leftrightarrow \forall e \in P : f_P \leq f_e^{\max} \quad [\text{as } f_e = f_P \; \forall e \in P] \\
&\Leftrightarrow f_P \leq -\hat{\omega}_P \\
&\Leftrightarrow \hat{f}_P \geq \hat{\omega}_P.
\end{aligned}
$$

Part (c)

$$
\begin{aligned}
(***) \; \exists e \in P : f_e > f_e^{\max} &\Leftrightarrow \forall e \in P : f_e > f_e^{\max} \\
&\qquad [\text{``}\Rightarrow\text{''} \text{ holds as } \forall e_1, e_2 \in P : f_{e_1} = f_{e_2} \\
&\qquad\qquad\qquad\qquad\qquad \wedge f_{e_1}^{\max} = f_{e_2}^{\max}] \\
&\Leftrightarrow f_P > f_e^{\max} \Leftrightarrow f_P > -\hat{\omega}_P \\
&\Leftrightarrow \hat{f}_P < \hat{\omega}_P.
\end{aligned}
$$

4.4 The Generalized Line Planning Game as an Instance of Games on Polyhedra

4.4.1 Formulation as a Game on a Polyhedron

In this section, we study the feasible frequency vectors of a line planning game. It turns out that the feasible region $\mathbb{F}^{\mathrm{LPG}}$ is a polyhedron and hence we can model the LPG as a game on a polyhedron. Games on polyhedra are generalized Nash equilibrium (GNE) games (see Chapter 3.4.3). Consequently, the line planning game is called a *generalized line planning game* if only feasible frequency vectors $f \in \mathbb{F}^{\mathrm{LPG}}$ are considered. In GNE games, the strategy set of each player depends on the strategies the other players choose.

Definition 4.11. *In generalized line planning games, a feasible frequency vector f^* is a* generalized equilibrium *if and only if for all lines $P \in \mathcal{P}$ and for all*

$$f_P \in [d_P^1(f^*_{-P}), d_P^2(f^*_{-P})] \cap \mathbb{R}_+,$$

it holds that

$$b_P(f^*_{-P}, f^*_P) \leq b_P(f^*_{-P}, f_P).$$

Generalized LPGs provide the advantage that each frequency vector and thus each generalized equilibrium is feasible. Furthermore, the payoff is given by $b_P(f) = c_P(f)$ The feasible region of an LPG

$$\mathbb{F}^{\mathrm{LPG}} = \left\{ f \in \mathbb{R}_+^{|\mathcal{P}|} : \sum_{P \in \mathcal{P}} f_P \geq f^{\min} \wedge \sum_{P : e \in P} f_P \leq f_e^{\max} \ \forall \, e \in E \right\}.$$

is a polyhedron, whose description is given next.

Definition 4.12. *In a network $G = (V, E)$ and for a path set $\mathcal{P}$, the* edge path incidence matrix H *has dimension $|E| \times |\mathcal{P}|$ and is given by the entries:*

$$h_{e,P} = \begin{cases} 1 & if \quad e \in P \\ 0 & else \end{cases}. \tag{4.7}$$

The edge path incidence matrix allows the following reformulation.

$$f_e = \sum_{P : e \in P} f_P = \sum_{P \in \mathcal{P}} h_{e,P} f_P = H_{e\cdot} f, \tag{4.8}$$

with $H_{e\cdot}$ being the row in H associated with edge e. Analogously, $H_{\cdot P}$ is the column in H that corresponds to line P.

$$c_P(f) = \sum_{e \in P} c_e(f_e) = \sum_{e \in E} c_e(f_e) h_{e,P} = (c_{e_1}(f_{e_1}), \ldots, c_{e_m}(f_{e_m})) H_{\cdot P}, \tag{4.9}$$

with $m = |E|$. Taking (4.8) and (4.9) together, we obtain:

$$c_P(f) = \sum_{e \in E} c_e \left(\sum_{P \in \mathcal{P}} h_{e,P} f_P \right) h_{e,P} = \sum_{e \in E} c_e \, (H_e.f) h_{e,P} = (c\,(Hf))^T \, H_{.P}.$$

$$(4.10)$$

We express constraints (4.2) and (4.3) by the following vector products.

$$\sum_{P \in \mathcal{P}} f_P = \mathbf{1}_{|\mathcal{P}|} f \geq f^{\min},$$

where $\mathbf{1}_n = (1, \ldots, 1)$ is the vector containing n times the entry 1, and

$$f_e = H_e.f \leq f_e^{\max} \; \forall \, e \in E.$$

The nonnegativity of the frequencies f_P can be expressed by using the identity matrix $\mathbb{I}_{|\mathcal{P}|}$.

$$\mathbb{I}_{|\mathcal{P}|} f \geq \mathbf{0}_{|\mathcal{P}|}.$$

Let $f^{\max} = (f_e^{\max})_{e \in E}$ be the vector of maximum frequencies, and let $\mathbf{0}_n = (0, \ldots, 0)^T$ be the vector containing n times the entry 0.

Summarizing, we get a representation of $\mathbb{F}^{\mathrm{LPG}}$ by $S(A, b) = \{f : Af \leq b\}$ with

$$A = \begin{pmatrix} -\mathbf{1}_{|\mathcal{P}|} \\ \hline H \\ \hline -\mathbb{I}_{|\mathcal{P}|} \end{pmatrix} \qquad b = \begin{pmatrix} -f^{\min} \\ \hline f^{\max} \\ \hline \mathbf{0}_{|\mathcal{P}|} \end{pmatrix}.$$

The matrix A is of dimension $(1 + |E| + |\mathcal{P}|) \times |\mathcal{P}|$. Hence, the generalized LPG is an instance of games on polyhedra. Note, that in line planning games, the payoff is minimized, whereas in games on polyhedra, we are maximizing payoff. This difference is not critical, as $\min b_P(f)$ is solved by the same frequency vectors f as $\max(-b_P(f))$, yielding the negative objective value. The next statement helps to transfer results from games on polyhedra to line planning games.

Lemma 4.13. *The polyhedron $S(A, b)$ representing an LPG with feasible region $\mathbb{F}^{\mathrm{LPG}}$ is compact.*

Proof. The polyhedron $S(A, b)$ is closed as it is given by $Af \leq b$. Furthermore, for all P and for all f_{-P} with $(f_{-P}, 0) \in \mathbb{F}^{\mathrm{LPG}}$, f_P lies in $[d_P^1, d_P^2] \cap \mathbb{R}_+$. As d_P^2 is bounded from above by $\min_{e \in P} f_e^{\max}$ it holds that $S(A, b)$ is bounded, and the claim follows. $\qquad\square$

Example 4.14. Consider the line planning game of Example 4.6 on page 121. The corresponding matrix H is given by

$$H = \begin{pmatrix} 1 & 0 \\ 0 & 1 \\ 1 & 1 \end{pmatrix}.$$

The polyhedron $S(A, b)$ is described by

$$A = \begin{pmatrix} -1 & -1 \\ 1 & 0 \\ 0 & 1 \\ 1 & 1 \\ -1 & 0 \\ 0 & -1 \end{pmatrix}, \quad b = \begin{pmatrix} -1 \\ 2 \\ 2 \\ 3 \\ 0 \\ 0 \end{pmatrix}.$$

4.4.2 Results of Using Polyhedric Representation

As we have an instance of games on polyhedra, we can transfer results from this type of games to the generalized line planning game. We show in this section that exact restricted potential functions exist for generalized line planning games. We propose a method to determine an equilibrium, valid for all types of cost functions. For linear and for strictly increasing costs c_e, we present a necessary and sufficient condition for equilibria. For strictly convex payoffs, we can show that all equilibria lie on the boundary of the feasible region $\mathbb{F}^{\mathrm{LPG}}$. The results obtained for generalized line planning games hold also for the standard line planning game, if only feasible frequencies are considered.

Lemma 4.15. *A generalized line planning game is a game on a polyhedron satisfying the PPG property (Definition 3.35).*

Proof. Consider a game with n players. Recall Definition 2.5: for a set of players $m \in \mathbb{P}(n)$, the class of edges is given by $e_m = \{e : \{P : e \in P\} = m\}$. We modify the network topology to obtain the standard network $G(n)$. All edges $e \in e_m$ that are shared by the same set of lines m have to be merged to a new single edge $\hat{e}_m$. It holds:

$$f_{\hat{e}_m} = \sum_{P : \hat{e}_m \in P} f_P = \sum_{P \in m} f_P = f_e \; \forall \, e \in e_m.$$

The costs of the new edges are given by

$$c_{\hat{e}_m}(f_{\hat{e}_m}) = \begin{cases} \sum_{e \in e_m} c_e(f_{\hat{e}_m}) & \text{if } e_m \neq \emptyset \\ 0 & \text{if } e_m = \emptyset \end{cases}.$$

By Lemma 2.15, the cost c_P of the line P, as a sum of edges contained in that line, stays invariant under this modification:

$$c_P(f) = \sum_{\hat{e}_m \in P} c_{\hat{e}_m}(f_{\hat{e}_m}) = \sum_{\hat{e}_m \in P} c_{\hat{e}_m} \left(\sum_{e \in e_m} c_e(f_{\hat{e}_m}) \right) = \sum_{e \in P} c_e(f_e).$$

The resulting line planning game is equivalent to the given one, as the modification only harms the structure of the network topology, whereas we are interested in the cost structure.

Let $n = |\mathcal{P}|$ be the number of lines and $\mathbb{P}(n)$ be the power set of the line pool $\mathcal{P}$. We transform the cost of a player P such that the PPG property (see Definition 3.35 on page 108) can be observed.

$$c_P(f) = \sum_{e \in P} c_e \left(\sum_{P_k : e \in P_k} f_{P_k} \right) = \sum_{m \in \mathbb{P}(n) : P \in m}' c_{\hat{e}_m}(f_{\hat{e}_m}). \tag{4.11}$$

$\square$

We use this insight for the following statements.

Theorem 4.16. *A generalized line planning game is an exact restricted potential game. Furthermore, equilibria exist in such a game.*

Proof. By Lemma 4.15, a generalized line planning game is a game on a polyhedron with the PPG property. By Lemma 3.40, such a game is an exact restricted potential game. Because by Lemma 4.13 we have a compact polyhedron, we conclude by using Lemma 3.41 that equilibria exist for these games. $\square$

Furthermore, the PPG property of a generalized line planning game enables the computation of generalized equilibria by using the following statement.

Theorem 4.17 (Method 1). *For a generalized line planning game with feasible region $\mathbb{F}^{\text{LPG}}$, a generalized equilibrium is given by an optimal solution of the following problem.*

$$\min \sum_{e \in E} [c_e(f_e) - c_e(0)] \qquad \text{subject to } f \in \mathbb{F}^{\text{LPG}}.$$

Proof. We use the idea of transformation (4.11) and rewrite:

$$\min \sum_{e \in E} [c_e(f_e) - c_e(0)] = -\max \left[-\sum_{m \in \mathbb{P}(n)} [c_{\hat{e}_m}(f_{\hat{e}_m}) - c_{\hat{e}_m}(0)] \right].$$

As a generalized line planning game can be represented by a game on a polyhedron $S(A, b)$ satisfying the PPG property (Lemma 4.15), the claim follows by Theorem 3.44. $\square$

Example 4.18. Consider the line planning game analyzed in Example 4.6. From $c_e(0) = 0 \ \forall \ e \in E$ we obtain $\sum_{e \in E} [c_e(f_e) - c_e(0)] = \sum_{e \in E} [c_e(f_e)]$. By Theorem 4.17 an equilibrium can be found by solving the following problem.

$$\min f_1 + 2f_2 + (f_1 + f_2)^2 \qquad \text{subject to } f \in \mathbb{F}^{\text{LPG}}. \tag{4.12}$$

The solution of the optimization problem, and thus an equilibrium is given by $f^* = (1, 0)$ with $b(f^*) = (2, 1)$. Note that f^* is the unique solution of (4.12), but not the unique equilibrium.

The example illustrates that Theorem 4.17 does not necessarily describe all equilibria of a game. For linear costs (and later also for strictly increasing costs), we present the following approach to determine all equilibria of the game. For linear cost functions c_e, the set of feasible equilibria is given by the solution set of an optimization problem.

Theorem 4.19 (Method 2). *Consider a generalized line planning game with feasible region* $\mathbb{F}^{\mathrm{LPG}}$ *and with linear cost functions*

$$c_e(f_e) = c_e \left(\sum_{P:e\in P} f_P \right) = a_e \sum_{P:e\in P} f_P, \ a_e \in \mathbb{R}.$$

Let $\alpha_{P,P_k} = \sum_{e\in P} a_e h_{e,P_k}$. *Furthermore, consider the following linear program.*

$$\min \sum_{P\in\mathcal{P}} sgn\,(\alpha_{P,P})\,\lambda_P f_P \qquad \text{subject to } f \in \mathbb{F}^{\mathrm{LPG}}. \qquad (4.13)$$

In this game, the following holds.

(a) If the frequency vector f^* *is an equilibrium, then there exists a vector* $\lambda \in \mathbb{R}^n_+$ *that satisfies* $\lambda_P > 0 \ \forall \ P$ *with* $\alpha_{P,P} \neq 0$ *such that* f^* *solves (4.13).*

(b) For all $\lambda \in \mathbb{R}^n_+$ *that satisfy* $\lambda_P > 0 \ \forall \ P$ *with* $\alpha_{P,P} \neq 0$, *the optimal solution of (4.13) is an equilibrium.*

Proof. In a line planning game it holds: $c_e(f_e) = c_e \left(\sum_{P:e\in P} f_P \right)$. Due to the linearity of c_e, we have

$$c_e(f_e) = a_e \sum_{P:e\in P} f_P, \ a_e \in \mathbb{R}.$$

We neglect the trivial case, where $a_e = 0$ holds $\forall \ e \in P$ and $\forall \ P \in \mathcal{P}$, because for it, each feasible $f \in \mathbb{F}^{\mathrm{LPG}}$ is an equilibrium and also a solution of (4.13).

Consider the payoff function of a player P:

$$c_P(f) = \sum_{e\in P} c_e(f_e) = \sum_{e\in P} \left(a_e \sum_{P_k\in\mathcal{P}:e\in P_k} f_{P_k} \right) = \sum_{e\in P} \left(a_e \sum_{P_k\in\mathcal{P}} h_{e,P_k} f_{P_k} \right)$$

$$= \sum_{e\in P} \left(\sum_{P_k\in\mathcal{P}} a_e h_{e,P_k} f_{P_k} \right) = \sum_{P_k\in\mathcal{P}} \left(\sum_{e\in P} a_e h_{e,P_k} f_{P_k} \right)$$

$$= \sum_{P_k\in\mathcal{P}} \left(f_{P_k} \sum_{e\in P} a_e h_{e,P_k} \right).$$

Hence, $c_P(f)$ is a linear function of the form

$$c_P(f) = \sum_{P_k\in\mathcal{P}} \alpha_{P,P_k} f_{P_k},$$

with $\alpha_{P,P_k} = \sum_{e \in P} a_e h_{e,P_k}$.

As there is a P such that $\exists\, e \in P : a_e \neq 0$, it holds that there is a P with $\alpha_{P,P} \neq 0$. Because the line planning game is an instance of games on polyhedra, and in addition, $\mathbb{F}^{\mathrm{LPG}}$ is compact for line planning games (see Lemma 4.13), the considered LPG satisfies the conditions of Theorem 3.6 (page 92). As the payoff in the LPG is going to be minimized (instead of being maximized such as in games on polyhedra), the claim follows. □

By Method 2, the set of equilibria can be found by solving (4.13) for all $\lambda \in \mathbb{R}^n_+$ that satisfy $\lambda_P > 0\ \forall\ P$ with $\alpha_{P,P} \neq 0$. The implementation of that approach will need an analytical study of that solution, and is not discussed here.

Example 4.20. Consider the line planning game analyzed in Example 4.6. We modify the cost functions to achieve linearity: $c_{e_1}(x) = x$, $c_{e_2}(x) = 2x$, and $c_{e_3}(x) = -3x$; see Figure 4.7.

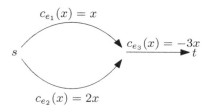

Fig. 4.7. Game network of Example 4.20.

We apply Theorem 4.19 and have $\alpha_{1,1} = 1 - 3 = -2$ and $\alpha_{2,2} = 2 - 3 = -1$ with $\mathrm{sgn}(\alpha_{1,1}) = \mathrm{sgn}(\alpha_{2,2}) = -1$. Thus, a frequency vector f^* is an equilibrium, if and only if it is a solution for some $\lambda_1, \lambda_2 > 0$ of the following linear program,

$$\min -\lambda_1 f_1 - \lambda_2 f_2 \qquad \text{subject to } f \in \mathbb{F}^{\mathrm{LPG}}.$$

We find the solution set of this problem and thus, the set of equilibria in the line planning game as $\{f = (f_1, 3 - f_1) : 1 \leq f_1 \leq 2\}$. For costs $c_3(x) = -2x$, we receive $\alpha_{1,1} = -1$ and $\alpha_{2,2} = 0$. In this case player 2 is indifferent about his flow f_2, whereas player 1 is trying to get as much flow as possible. Thus the set of equilibria is then given by $\{f = (2, f_2) : 0 \leq f_2 \leq 1\}$. For $c_3(x) = -3/2x$, we obtain a unique equilibrium: $\alpha_{1,1} = -1/2$ and $\alpha_{2,2} = 1/2$ delivers the equilibrium $f = (2, 0)$.

For strictly increasing costs we obtain a related result.

Theorem 4.21 (Method 3). *Consider a generalized line planning game with feasible region $\mathbb{F}^{\mathrm{LPG}}$, where the payoffs c_e are strictly increasing in f_e. Furthermore, consider the following linear problem.*

$$\min \sum_{P \in \mathcal{P}} \lambda_P f_P \qquad subject\ to\ f \in \mathbb{F}^{\mathrm{LPG}}. \tag{4.14}$$

In this game, the following hold.

(a) If the frequency vector f^* is an equilibrium, then there exist $\lambda_P > 0, P \in \mathcal{P}$, such that f^* solves (4.14).

(b) For all $\lambda_P > 0, P \in \mathcal{P}$, the optimal solution f^* of (4.14) is an equilibrium.

Proof. From strictly increasing c_e, we obtain $c_P(f) = \sum_{e \in P} c_e(f_e)$ strictly increasing in f_P (see also Proposition 2.44). As the line planning game is an instance of games on polyhedra, and in addition, $\mathbb{F}^{\mathrm{LPG}}$ is compact for line planning games (see Lemma 4.13), the considered LPG satisfies the conditions of Theorem 3.9. As in the LPG the payoff is minimized, the claim follows.   □

By Method 3, the complete set of equilibria can be found by solving (4.13) for all $\lambda_P > 0, P \in \mathcal{P}$. The implementation of that approach will need an analytical study of that solution, and is not discussed here.

Example 4.22. Consider the line planning game analyzed in Example 4.6. We have strictly increasing cost functions c_e in this instance. Hence, we can apply Theorem 4.21. A frequency vector f^* is an equilibrium in this example, if and only if it is a solution of the following linear program for $\lambda_1, \lambda_2 > 0$,

$$\min \lambda_1 f_1 + \lambda_2 f_2 \qquad subject\ to\ f \in \mathbb{F}^{\mathrm{LPG}}.$$

We find the solution set of this problem and thus, the set of equilibria in the presented line planning game as $\{f = (f_1, 1 - f_1) : 0 \le f_1 \le 1\}$.

The next result holds for line planning games with convex costs.

Lemma 4.23. *In a generalized line planning game with strictly convex cost functions c_e, all equilibria lie on the boundary of the feasible region $\mathbb{F}^{\mathrm{LPG}}$.*

Proof. As we have strictly convex cost functions c_e on the edges, the cost functions $c_P(f)$ for the players are also strictly convex (see Lemma 2.18 on page 20 and note that it also holds for the case of strict convexity). As we have existence of equilibria ensured by Theorem 4.16, the claim follows by Lemma 3.13 (page 98).   □

Comparison of Methods 1–3

	Positive	Negative
Method 1 (Theorem 4.17)	Can be used for any cost function	All equilibria are not necessarily found
Method 2 (Theorem 4.19)	Provides complete set of equilibria	Only valid for linear costs
Method 3 (Theorem 4.21)	Provides complete set of equilibria	Only valid for strictly increasing costs

Depending on the type of cost function, one of the described methods is to be chosen. For line planning games, strictly increasing costs are quite usual. In these cases, Method 3 is a suitable tool that is able to find all equilibria of a line planning game.

4.5 Extensions of the Line Planning Game

4.5.1 Integer Line Planning Game

In terms of the practical application of line planning games, frequencies f_P represented by real numbers are not acceptable, as in real-world problems, frequencies have to be natural numbers. In this section, we extend the basic line planning game to an *integer line planning game (ILPG)*, where $f_P \in \mathbb{N}_0$ is required. In our attempt to obtain feasible solutions, we consider in particular the *generalized integer line planning game*, where we focus on feasible frequencies. Formally, the generalized integer line planning game is an integer version of generalized Nash equilibria (GNE) games (see page 85). To our knowledge, integer GNE games have not been studied yet and it would be interesting to investigate this field in general and in particular for integer games on polyhedra. Then, the *set of feasible frequency vectors* is given by

$$\mathbb{F}^{\text{ILPG}} = \left\{ f : f_P \in \mathbb{N}_0 \ \forall P \in \mathcal{P} \land \sum_{P \in \mathcal{P}} f_P \geq f^{\min} \land \sum_{P : e \in P} f_P \leq f_e^{\max} \ \forall e \in E \right\}.$$

All other components of the model, such as the cost functions, and lower and upper bounds, are equal to those definitions of the line planning game treated in previous sections. The definition of an equilibrium, however, needs a revision for integer line planning games.

Definition 4.24. *In an integer line planning game, a frequency vector f^* is an equilibrium if and only if for all lines $P \in \mathcal{P}$ and for all $f_P \in \mathbb{N}_0$ it holds that*

$$b_P(f^*_{-P}, f^*_P) \leq b_P(f^*_{-P}, f_P).$$

Example 4.25. Consider the line planning game presented in Example 4.6 with minimal frequency $f^{\min} = 1$ and maximal frequencies $f_{e_1}^{\max} = f_{e_2}^{\max} = 2$ and $f_{e_3}^{\max} = 3$. In the extension to an integer line planning problem, we obtain the set of feasible frequencies $\mathbb{F}^{\text{ILPG}}$ as illustrated in Figure 4.8.

The ILPG is in contrast to line planning games (and also to path player games and games on polyhedra) a finite game. As in line planning games, we focus in our research on feasible frequencies, as infeasible solutions are not implementable in practice. Thus, we consider *generalized integer line planning games*.

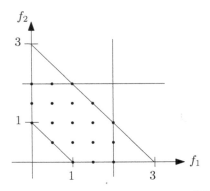

Fig. 4.8. Set of feasible frequencies $\mathbb{F}^{\text{ILPG}}$.

Definition 4.26. *In a generalized integer line planning game, a frequency vector f^* is a* generalized equilibrium *if and only if for all lines $P \in \mathcal{P}$ and for all $f_P \in \mathbb{F}^{\text{ILPG}}$ it holds that*

$$b_P(f^*_{-P}, f^*_P) \le b_P(f^*_{-P}, f_P).$$

Theorem 4.27. *A generalized integer line planning game is an exact restricted potential game.*

Proof. By Theorem 4.16, generalized line planning games are exact restricted potential games. That means we can find an exact restricted potential function $\Pi(f)$ such that for all f_{-P} with $(f_{-P}, 0) \in \mathbb{F}^{\text{LPG}}$ and for every $x, z \in [d^1_P, d^2_P]$ it holds:

$$b_P(f_{-P}, x) - b_P(f_{-P}, z) = \Pi(f_{-P}, x) - \Pi(f_{-P}, z). \qquad (4.15)$$

As

$$\mathbb{F}^{\text{ILPG}} \subseteq F^{\text{LPG}} \quad \text{and} \quad \left(\mathbb{N}_0 \cap [d^1_P, d^2_P]\right) \subseteq \left(\mathbb{R}_+ \cap [d^1_P, d^2_P]\right),$$

it follows that (4.15) also holds for all f_{-P} with $(f_{-P}, 0) \in \mathbb{F}^{\text{ILPG}}$ and for every $x, z \in \left(\mathbb{N}_0 \cap [d^1_P, d^2_P]\right)$. $\qquad \square$

Theorem 4.28. *In an integer line planning game, feasible equilibria exist if $\mathbb{F}^{\text{ILPG}}$ is nonempty.*

Proof. Consider the generalized integer line planning game with respect to $\mathbb{F}^{\text{ILPG}}$. As this game is an exact restricted potential game, equilibria are given by minimizers of the potential function. Because we have a finite number of frequencies f in $\mathbb{F}^{\text{ILPG}}$, minimizers exist and so do equilibria in the generalized integer line planning game. By the definition of the payoff function, these equilibria are also equilibria in the corresponding integer line planning game and they are feasible. $\qquad \square$

As the generalized ILPG is a finite exact restricted potential game, it satisfies the finite improvement property (see Definition 2.82 and Lemma 2.83 on page 62). As FIP implies the finite best-reply property (FBRP) (see Definition 2.110) we can obtain a feasible equilibrium in finite time by the procedure described in Algorithm 2. We use the following definition.

Definition 4.29. *In an integer LPG, the* best reaction set *for a line P is defined for fixed f_{-P} by*

$$f_P^{br} = \left\{ f_P \in \mathbb{N}_0 \ \cap \ [d_P^1, d_P^2] : f_P \text{ minimizes } b_P(f_{-P}, f_P) \right\}.$$

Note, that in Chapter 2, the notation $f_P^{\max}$ is used for the equivalent definition. To prevent confusion with $f_e^{\max}$, $f^{\min}$, we changed this term.

Algorithm 2 Best-reply improvement path for ILPG

1: Set initial frequency vector: $f^0 \in \mathbb{F}^{\text{ILPG}}$
2: **while** $\exists \, P \in \mathcal{P} : f_P \notin f_P^{br}$ **do**
3:　　Select $\bar{P} \in \mathcal{P} : f_{\bar{P}} \notin f_{\bar{P}}^{br}$
4:　　Select $\bar{f}_{\bar{P}} \in f_{\bar{P}}^{br}$
5:　　Set $f_{\bar{P}} = \bar{f}_{\bar{P}}$
6: **end while**
7: Frequency f is a feasible equilibrium.

Algorithm 2 starts with a feasible frequency vector. By following best-reply improvement steps, it will never produce an infeasible solution throughout the procedure. As long as there is a player that can improve her payoff, in each step such a player shifts to a best-reply strategy. As in each step, a potential function $\Pi(f)$ is also strictly increasing, no frequency vector f is visited twice. As the set of feasible frequencies is bounded, the algorithm is finite. This observation is provided by the finite best-reply improvement property. Hence, Algorithm 2 terminates after a finite number of steps, with a feasible equilibrium.

One major question in terms of implementation of this algorithm is the choice of an initial frequency vector $f^0 \in \mathbb{F}^{\text{ILPG}}$. Indeed, it may be difficult to find such a feasible integer solution, and it may even be unclear if $\mathbb{F}^{\text{ILPG}}$ is nonempty. A question left open for future research is, whether the special structure of the polyhedra defining the line planning game can be exploited to determine feasible starting frequencies. Furthermore, the choice of the initial frequency vector and the selection of the active player $\bar{P}$ influences the result of the algorithm. Thus, heuristic rules for these selections could also be part of further research in this area.

As we assume continuous cost functions, f_P^{br} is nonempty for all lines P (although it does not necessarily include an integer point); see Lemma 2.24 on page 22. For the determination of f_P^{br}, appropriate numerical algorithms are chosen. Depending on the type of cost function, line search algorithms or subgradient methods or other suitable methods may be used (see [BS79]).

4.5.2 Multiple Origin–Destination Pairs

We consider a network $G = (V, E)$ with Q multiple origin–destination(OD) pairs $\{s_q, t_q\}$, $q = 1, \ldots, Q$. For the qth OD pair, the pool of lines connecting s_q and t_q is given by $\mathcal{P}_q$. The paths are given such that we have pairwise disjoint sets:

$$\mathcal{P}_{q_1} \cap \mathcal{P}_{q_2} = \emptyset \ \forall \ q_1, q_2 = 1, \ldots, Q, \ q_1 \neq q_2.$$

With $q(P)$ we denote the index of the OD pair $\{s_q, t_q\}$ such that $P \in \mathcal{P}_q$. Because each line P is assigned to exactly one OD pair, $q(P)$ is well defined. Furthermore, the minimal frequency for the qth OD pair is given by $f_q^{\min}$. We denote:

$$\mathcal{P} = \bigcup_{q=1,\ldots,Q} \mathcal{P}_q \quad \text{and} \quad f^{\min} = \left(f_q^{\min} \right)_{q=1,\ldots,Q}.$$

The maximal frequencies on edges $f_e^{\max}$ and the cost c_e assigned to the edges are defined as in the single origin–destination case. We call such a game a *line planning game with multiple OD pairs*.

Definition 4.30. *The payoff for player P and a frequency vector $f \geq 0_{|\mathcal{P}|}$ in an LPG with multiple OD pairs is given by*

$$b_P(f) = \begin{cases} c_P(f) & \text{if} \sum_{P_k \in \mathcal{P}_{q(P)}} f_{P_k} \geq f_{q(P)}^{\min} \wedge \forall \ e \in P : f_e \leq f_e^{\max} \\ N & \text{if} \sum_{P_k \in \mathcal{P}_{q(P)}} f_{P_k} \geq f_{q(P)}^{\min} \wedge \exists \ e \in P : f_e > f_e^{\max} \\ M & \text{if} \sum_{P_k \in \mathcal{P}_{q(P)}} f_{P_k} < f_{q(P)}^{\min} \end{cases}.$$

As in the single OD pair case (see Definition 4.2), a frequency vector f is called *feasible* if the bounds $f_q^{\min}, q = 1, \ldots Q$ and $f_e^{\max}, e \in E$ are satisfied. The *set of feasible frequencies for line planning games with multiple OD pairs* is given by

$$\mathbb{F}^{\text{LPGMOD}} = \left\{ f \in \mathbb{R}_+^{|\mathcal{P}|} : \sum_{P \in \mathcal{P}_q} f_P \geq f_q^{\min} \ \forall q \in Q \wedge \sum_{P : e \in P} f_P \leq f_e^{\max} \ \forall e \in E \right\}.$$

Finally, we have to adjust the definition of the lower decision limit presented in (4.4):

$$d_P^1(f_{-P}) = f_{q(P)}^{\min} - \sum_{\substack{P_k \in \mathcal{P}_q \\ P_k \neq P}} f_{P_k},$$

whereas the upper decision limit stays the same, as presented in (4.5):

$$d_P^2(f_{-P}) = \min_{e \in P} \{f_e^{\max} - f_{e,-P}\}.$$

If we just consider feasible frequencies $f \in \mathbb{F}^{\text{LPGMOD}}$, we obtain a *generalized line planning game with multiple OD pairs*. Generalized equilibria are defined in such games similar to the single OD pair case in Definition 4.11.

Recall the definition of the edge path incidence matrix H (see Definition 4.12). A line planning game with multiple OD pairs is represented by a game on a polyhedron $S(A,b)$ with:

$$A = \begin{pmatrix} -\mathbf{1}_{|\mathcal{P}_1|} & 0 & \cdots & 0 & 0 \\ 0 & -\mathbf{1}_{|\mathcal{P}_2|} & \cdots & 0 & 0 \\ \vdots & \vdots & \ddots & \vdots & \vdots \\ 0 & 0 & \cdots & -\mathbf{1}_{|\mathcal{P}_{m-1}|} & 0 \\ 0 & 0 & \cdots & 0 & -\mathbf{1}_{|\mathcal{P}_m|} \\ \hline & & H & & \\ \hline & & -\mathbb{I}_{|\mathcal{P}|} & & \end{pmatrix} \qquad b = \begin{pmatrix} -f_1^{\min} \\ -f_2^{\min} \\ \vdots \\ -f_m^{\min} \\ \hline f^{\max} \\ \hline \mathbf{0}_{|\mathcal{P}|} \end{pmatrix}.$$

Example 4.31. We consider a line planning game with four OD pairs as illustrated in Figure 4.9. Let $f_q^{\min} = 1 \ \forall \ q = 1, \ldots, Q$ and $f_e^{\max} = 4 \ \forall \ e \in E$. We denote the edges $e = 1, \ldots, 24$ and the lines with $P^1, \ldots, P^{10}$. The frequency of the lines is given by $f_1, \ldots, f_{10}$.

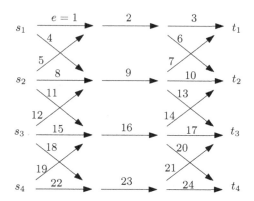

Fig. 4.9. Game network of Example 4.31.

The line pools are given by

$$\begin{aligned}
\mathcal{P}_1 &= \{P^1, P^2\} &&= \{(1,2,3),(4,9,7)\}, \\
\mathcal{P}_2 &= \{P^3, P^4, P^5\} &&= \{(5,2,6),(8,9,10),(11,16,14)\}, \\
\mathcal{P}_3 &= \{P^6, P^7, P^8\} &&= \{(12,9,13),(15,16,17),(18,23,21)\}, \\
\mathcal{P}_4 &= \{P^9, P^{10}\} &&= \{(19,16,20),(22,23,24)\}.
\end{aligned}$$

The polyhedron $S(A,b)$ representing $\mathbb{F}^{\mathrm{LPGMOD}}$ is described in Appendix A.2. We introduce cost functions $c_e(f_e) = f_e$ for all edges e in E. We apply Theorem 4.17 and solve

$$\min \sum_{e \in E} [c_e(f_e) - c_e(0)] = \min \sum_{e \in E} c_e(f_e)$$

$$= \min \sum_{e \in E} c_e \left(\sum_{P \in \mathcal{P}} h_{e,P} f_P \right)$$

$$= \min \left(3f_1 + 3f_2 + 3f_3 + 3f_4 + 3f_5 + 3f_6 + 3f_7 + 3f_8 + 3f_9 + 3f_{10} \right)$$

subject to $f \in S(A,b)$.

As each frequency f_P has exactly the same coefficient in the objective function, each frequency that satisfies $\sum_{P \in \mathcal{P}_q} f_P = f_q^{\min} = 1$, for example, $f^1 = (1,0,1,0,0,1,0,0,1,0)$, is an optimal solution and thus also an equilibrium. The objective value is 4 for all these solutions. The payoff for f^1 is given by $b(f^1) = (4,1,4,1,1,3,1,0,3,0)$, whereas, for example, for $f^2 = (1,0,0,1,0,0,1,0,0,1)$ we have a payoff $b(f^2) = (3,1,1,3,1,1,3,1,1,3)$.

We can use this approach also for nonlinear cost functions. Set, for instance, $c_e(f_e) = f_e^2$ for all edges e in E. The objective function $\min \sum_{e \in E} c_e(f_e)$ yields the optimal solution

$$f^3 = (0.538, 0.462, 0.385, 0.308, 0.308, 0.308, 0.308, 0.385, 0.462, 0.538)$$

with an objective value of 7.385. Solving this problem as an integer problem yields $f^4 = (0,1,1,0,0,0,0,1,1,0)$, with objective value 12.

4.6 Line Planning for Interregional Trains in Germany

The line planning game chapter closes with a numerical example. We implement one of our approaches, namely Method 1, Theorem 4.17, using real-world data related to the German railway system of Deutsche Bahn AG. In particular we consider train stations connected by interregional trains, such as InterCityExpress (ICE), InterCity (IC), and EuroCity (EC). The following studies are meant to test the possibility of implementing our method with realistic data and to obtain equilibria based on larger databases. Our numerical study is interesting for the following two reasons.

- Although the investigation of line planning games is still in an early stage, and the results are hence not ready for practical use yet, the study illustrates that further research in this field is worthwhile.
- Second, the numerical behavior of the methods developed for games on polyhedra is demonstrated.

The following data are at our disposal.

- OD matrix describing 319 train stations and the minimal frequency $f_q^{\min}$ given for the OD pairs
- Three line databases of different size, containing 132 (S, small), 688 (M, medium) and 2770 (L, large) lines

The line databases are not in a form suitable for our model. We discuss later how line pools are created from these data. From theoretically $319 \times 318 = 101,442$ OD pairs, $56,646$ still have a positive minimal frequency $f_q^{\min}$ and have to be considered. Thus, we have a line planning game with multiple OD pairs. For those OD pairs, $f_q^{\min} \in [1, 4831]$ hold. Note that the values of $f^{\min}$ are to be interpreted as weights dependent on the number of passengers. From these weights, frequencies are obtained by a linear transformation. The train stations under consideration are located in Germany and neighboring countries. Figure 4.10 illustrates the locations of all 319 stations.

The following information is needed for the line planning game, but is not provided by the data. We do not have available the maximal frequency $f_e^{\max}$ on the edges, nor do we know the costs c_e assigned to the edges. Thus, we have to make assumptions for the implementation of our model. As there is no maximal frequency on the edges, we choose the value sufficiently large for each edge, such that the maximal frequency is satisfied for our problems. In particular, we set $f_e^{\max} = 100,000$. Regarding the cost function, the strictly increasing function

$$c_e(f_e) = f_e$$

is implemented on all edges $e \in E$.

As the line planning game model considers only direct connections between stations, we neglect all OD pairs where no direct connection exists in the line pool. For the future design of line databases, this should be taken into consideration. Furthermore, we introduce the bound U_q and consider only OD pairs where $f_q^{\min} > U_q$ does hold. This bound is used to consider only OD pairs of strong influence; that is, with high minimal frequencies for our computations and it is a tool to control the size of the problem.

Furthermore, we have to construct a line pool from the line databases according to the definitions in our model. As we reduce the number of OD pairs, we have to analyze only lines that are relevant for the OD pairs under consideration. Thus, we generate the line pool by using these lines. On the other hand, one line may offer a direct connection for more than one OD pair. In our model, we assume disjoint line pools: one line has to be assigned to exactly one OD pair. According to this, we duplicate lines that provide a

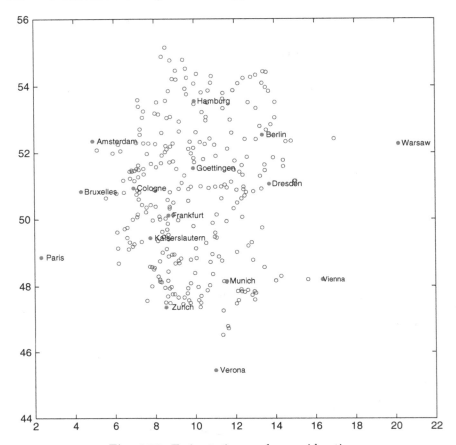

Fig. 4.10. Train stations under consideration

direct connection for more than one OD pair. The lines have to be given such that we obtain a line pool $\mathcal{P} = \bigcup_{q=1,\dots,Q} \mathcal{P}_q$ consisting of disjoint subsets $\mathcal{P}_q$. Note that the frequencies of the original lines from the databases S, M, and L are then given by the sum over the frequencies of the original lines' duplicates.

We study five scenarios with a different number of OD pairs and use different line databases. In Studies 1, 2, and 3, we consider the same set of OD pairs, namely for $f_q^{\min} > 599$, but we change the size of the line database. In Studies 2, 4, and 5, the line database is invariant (we choose the medium-sized one), but the set of OD pairs is changed.

	$f_q^{\min} > 999$	$f_q^{\min} > 599$	$f_q^{\min} > 399$
Small		Study 1	
Medium	Study 4	Study 2	Study 5
Large		Study 3	

Table 4.1 contains the computational results. We present a short explanation of the content in the following list.

Column 3. Number of OD pairs that satisfy $f_q^{\min} > U_q$

Column 5. Size of line databases

Column 6. Number of OD pairs with direct connections and that satisfy $f_q^{\min} > U_q$

Column 7. Size of line pool constructed from line database, including duplicates of lines

Column 8. Number of lines with positive frequency; that is, that are established for the PTN (including duplicates)

Column 9. Objective function value of the optimization problem solved with Method 1

Column 10. Reference to Figure of PTN

Columns 12–14. Copied from the first part of the table, for easier reading

Columns 15–19. Statistical information about length of each line (number of stations)

Columns 20–24. Statistical information about number of lines (including duplicates) serving each train station

Table 4.1. Computational results.

1	2	3	4	5	6	7	8	9	10
Study	U_q	# OD pairs	Line Data-base	Size Data-base	# OD pairs with Direct Connections	Size Line Pool	Lines with Positive Frequency	$c \times x$	PTN Figure
1	599	251	S	132	87	262	88	$1,402,494.001$	4.11
2	599	251	M	688	117	1287	156	$2,151,352.000$	4.12
3	599	251	L	2770	157	5544	244	$2,636,404.000$	4.13
4	999	113	M	688	53	493	68	$1,456,873,000$	4.14
5	399	499	M	688	132	2610	299	$2,971,507.012$	4.15

11	12	13	14	# of Stations per Line					# of Lines per Station				
				15	16	17	18	19	20	21	22	23	24
Study	Line Data-base	# OD pairs with Direct Connections	# of Chosen Lines	Min	Max	Mean	Var	Histogram Figure	Min	Max	Mean	Var	Histogram Figure
1	S	87	88	6	33	15.15	25.94	4.16	1	43	4.18	52.76	4.21
2	M	117	156	9	20	15.06	10.87	4.17	1	73	7.36	131.97	4.22
3	L	157	244	6	37	14.18	23.83	4.18	1	121	10.84	323.17	4.23
4	M	53	68	9	20	15.21	11.66	4.19	1	34	3.24	30.59	4.24
5	M	132	299	9	20	15.34	10.44	4.20	1	129	14.38	435.38	4.25

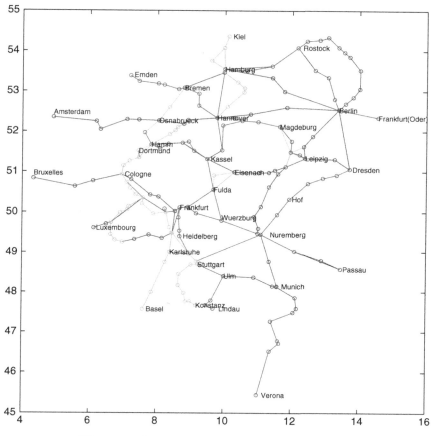

Fig. 4.11. PTN of Study 1, $U_q = 599$, line pool S.

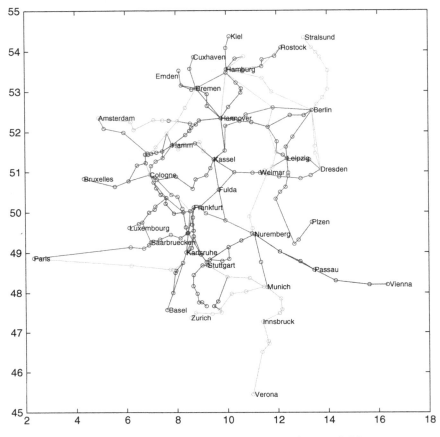

Fig. 4.12. PTN of Study 2, $U_q = 599$, line pool M.

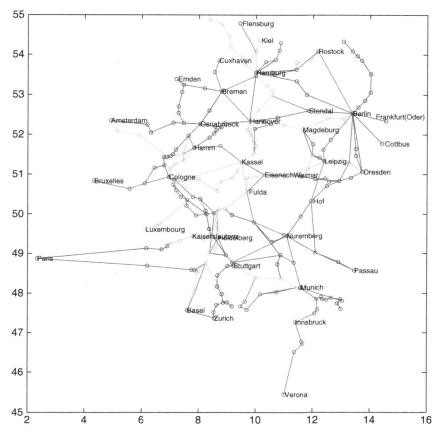

Fig. 4.13. PTN of Study 3, $U_q = 599$, line pool L.

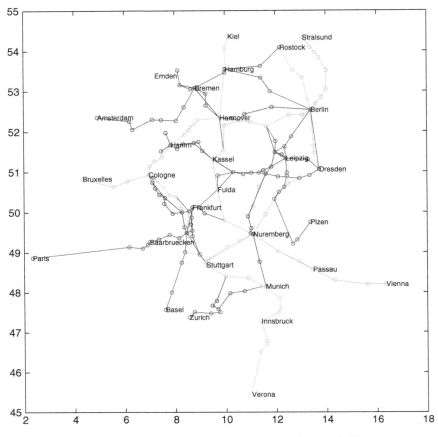

Fig. 4.14. PTN of Study 4, $U_q = 999$, line pool M.

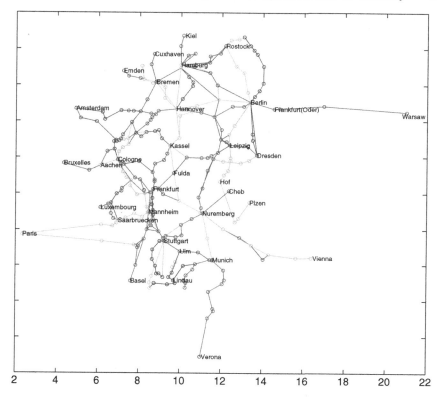

Fig. 4.15. PTN of Study 5, $U_q = 399$, line pool M.

Histogram: Number of Stations per Line

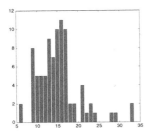

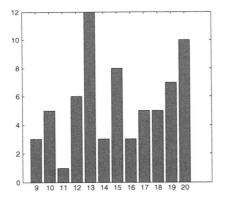

Fig. 4.16. Study 1, $U_q =$ 599, S.

Fig. 4.17. Study 2, $U_q =$ 599, M.

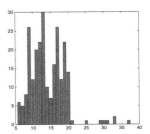

Fig. 4.18. Study 3, $U_q =$ 599, L.

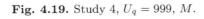

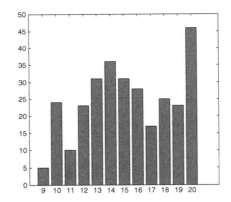

Fig. 4.19. Study 4, $U_q = 999$, M.

Fig. 4.20. Study 5, $U_q = 399$, M.

Histogram: Number of Lines per Station

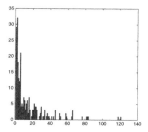

Fig. 4.21. Study 1, $U_q = 599$, S.

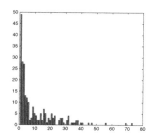

Fig. 4.22. Study 2, $U_q = 599$, M.

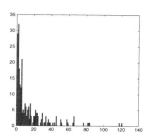

Fig. 4.23. Study 3, $U_q = 599$, L.

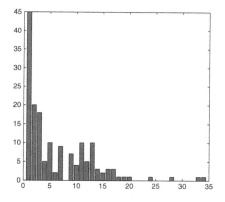

Fig. 4.24. Study 4, $U_q = 999$, M.

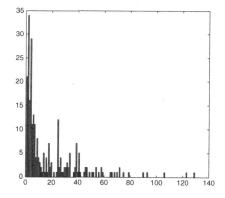

Fig. 4.25. Study 5, $U_q = 399$, M.

In terms of lines per station, the station served by the highest number of lines in each study is Frankfurt(Main) Süd. For each study, the 12 most frequented stations in descending order are as follows.

Study 1: $U_q = 599$, S

1. Frankfurt(M) Süd
2. Hanau
3. Fulda
4. Mannheim
5. Kassel-Wilhelmshöhe
6. Stuttgart
7. Vaihingen
8. Köln-Deutz
9. Nürnberg
10. Göttingen
11. Frankfurt Airport
12. Hannover

Study 2: $U_q = 599$, M

1. Frankfurt(M) Süd
2. Hanau
3. Fulda
4. Hannover
5. Kassel-Wilhelmshöhe
6. Göttingen
7. Hamburg Hbf
8. Stuttgart
9. HH-Harburg
10. Nürnberg
11. Vaihingen
12. Köln-Deutz

Study 3: $U_q = 599$, L

1. Frankfurt(M) Süd
2. Hanau
3. Fulda
4. Hannover
5. Kassel-Wilhelmshöhe
6. Göttingen
7. Hamburg Hbf
8. HH-Harburg
9. Köln-Deutz
10. Nürnberg
11. Berlin Lehrter
12. Frankfurt Hbf

Study 4: $U_q = 999$, M

1. Frankfurt(M) Süd
2. Hanau
3. Fulda
4. Hannover
5. Hamburg Hbf
6. Kassel-Wilhelmshöhe
7. HH-Harburg
8. Hamm (Westf.)
9. Nürnberg
10. Köln-Deutz
11. HH Dammtor
12. Frankfurt Hbf

Study 5: $U_q = 399$, M

1. Frankfurt(M) Süd
2. Hanau
3. Fulda
4. Köln-Deutz
5. Hannover
6. Kassel-Wilhelmshöhe
7. Frankfurt Hbf
8. Nürnberg
9. Hamburg Hbf
10. HH-Harburg
11. Stuttgart
12. Göttingen

5

Summary

In this work a new type of network game, the *path player game* has been introduced and analyzed. In this game, players are represented by paths in a network. Hence, it models network flows from the point of view of the network operators if we assume that the paths are owned by competing individuals. For a future extension of the model, it will be worthwhile to consider not only paths, but complete subnetworks as players. For this investigation, it makes sense to distinguish the general case from that case, where each player–subnetwork contains source and sink. In the scope of this research, the introduction of multiple sources and sinks will be interesting. The path player game belongs to the class of network games and is related to the routing game [KP99, CSS04b, Rou05a], to the bandwidth allocation game [Kel97, JT04], and to path auctions [ESS04, AT02]. We have been able to prove the existence of pure-strategy equilibria in path player games, which is a nontrivial task due to the noncontinuous payoff functions and the infinite, mutually dependent strategy sets. Moreover, we analyzed different instances of path player games in more detail and presented characterizations for these cases. For instance, we introduced the noncompensative security (NCS) property, which provides together with strictly increasing costs a necessary and sufficient condition for a profile of flows to be in equilibrium.

Path player games may have multiple equilibria. This observation motivated the analysis of the relation of equilibria and nondominated solutions in the sense of Pareto (see, e.g., [Ehr05]). It turned out that in fact all relations between the set of nondominated solutions and the set of equilibria are possible. In particular, an example related to the Prisoner's Dilemma (see, e.g., [Owe95]) has been presented, where each equilibrium is dominated and each nondominated solution is nonstable. But also classes of games with nice behavior have been described, where the set of equilibria and the set of nondominated solutions are equal or, at least, each nondominated solution is an equilibrium.

The existence of pure-strategy equilibria in path player games motivated the investigation of potential functions [Ros73, MS96]. We started the analysis

S. Schwarze, *Path Player Games*, DOI 10.1007/978-0-387-77928-7_5,
© Springer Science+Business Media, LLC 2009

by considering only feasible flows, which led to mutual dependent strategy sets. We introduced the new definition of restricted potential functions, which takes the dependencies of strategy sets into account. We were able to show the existence of exact restricted potential functions for path player games. From this result a second proof for the existence of equilibria and a method to compute equilibria by maximizing the potential function was derived. Furthermore, we proved the existence of an ordinal potential function, a weaker type of a potential function, if we consider the original path player game, also allowing infeasible flows. This result was strengthened in a second approach to the existence of exact potential functions by extending the payoff function. Apart from maximizing the potential function, a second approach for computation of equilibria was derived from the existence of potential functions. Best-reply improvement sequences were investigated, which, if they are maximal, end up with equilibria. We were able to describe instances where the best-reply improvement sequences are finite, although we studied a problem with an infinite number of strategies. For all other cases, we showed that approximate equilibria can be obtained using ε-improvement sequences. For further research a more detailed analysis of the improvement sequence algorithm would be interesting. Which initial solution should be chosen? How should the active player be chosen? These decisions have to be taken before implementing the algorithm and will influence the performance of the procedure. Moreover, how can the attraction region of an equilibrium solution be described (i.e., which set of initial solutions will lead to a particular equilibrium) when using an improvement sequence?

We have extended the concept of path player games to a network-independent approach, the newly introduced games on polyhedra. These games are a special instance of generalized Nash equilibrium games [AD54, Har91, FP06] and thus are also highly interesting from a theoretical point of view. The crucial property of generalized Nash equilibrium games is that the players' strategy sets are not fixed, but mutually dependent on each other. In games on polyhedra, these dependencies are described by a polyhedron. The existence of equilibria is not given in general for games on polyhedra, which we demonstrated by an example. Nevertheless, for strictly increasing payoffs or for linear payoffs, the existence of equilibria was proven (for bounded polyhedra) and even more, it is possible to determine the complete set of equilibria by solving a linearly restricted optimization problem. For linear payoffs or if the so-called path player game property is satisfied, games on polyhedra have restricted potential functions. In these cases, the existence of equilibria is given for bounded polyhedra and the computation of equilibria is possible by maximizing the potential function. For the investigation of games on polyhedra we have introduced the game on a hypercuboid, a tool that allows the analysis without the mutual dependency of strategy sets. To this end, we analyzed the situation for the smallest hypercuboid that contains the polyhedron of interest. For future research in the field of games on polyhedra, an integer version of the problem is an interesting next step, which also makes sense with respect to some applications, such as the line planning problem. For such an integer version, the improvement sequence approach for games with potential functions is highly interesting, as we can assume a finite number of solutions for a bounded polyhedron. Also here it will be interesting how the initial solution is chosen, and the question arises if such an initial solution does exist.

Finally, we presented the line planning game, which applies the results of the previous chapters to a problem from transport optimization, the line planning problem. It turned out that the line planning game is an instance of games on polyhedra and in particular it has the path player game property. Thus, the existence of potential functions was given. This implied the existence of equilibria if the polyhedron of the line planning game was nonempty, as the polyhedron is bounded by the nature of the line planning game. Based on these results, three algorithmic approaches for computation of equilibria were presented. The first method maximizes the potential function and will not necessarily find all equilibria, whereas the other two methods are valid only for strictly increasing or linear payoffs, but yield the complete set of equilibria. We discussed a multisource-multisink version of the game which is more appropriate for realistic problems. Moreover, because the line planning problem is integer, we studied an integer version of the line planning game and proposed an algorithm for determining equilibria based on the best-reply improvement sequence. To test the numerical behavior of our first method in a realistic setting and to determine equilibria with real-world data, we implemented the maximization of the potential function. For this purpose, we used data from the Deutsche Bahn AG, in particular from the German interregional trains. These first results are a motivation to continue the research in this area. Of course, the model is in an early stage. The following extensions would be first steps towards a more realistic setting. First of all the introduction of setup costs for establishing lines is a realistic assumption. Moreover, in our approach, we minimize the expected delay of a line. In addition, the consideration of the length (e.g., travel time) of a line, maybe even in a multicriteria approach would be meaningful. Also, the model currently considers only direct connections from source to sink. This is not satisfying and it would be nice to allow passengers to change lines. Additional implementions and numerical tests will be important to check the suitability of the model.

A

Appendix

A.1 Lemma: Transformation of Line Planning Game

Lemma A.1 *Consider a line planning game $\Gamma = (G, \mathcal{P}, f^{\min}, f^{\max}, c, N, M)$ with cost functions $c_e(f_e) \leq 0$ for $f_e \leq 0$. Furthermore, we have $f^{\min} \leq 0$ and $f_e^{\max} \leq 0, e \in E$. We obtain a transformed line planning game $\bar{\Gamma} = (G, \mathcal{P}, \bar{f}^{\min}, \bar{f}^{\max}, \bar{c}, N, M)$ with $\bar{c}_e(\bar{f}_e) \geq 0$ for $\bar{f}_e \geq 0$, $\bar{f}^{\min} \geq 0$ and $\bar{f}_e^{\max} \geq 0, e \in E$ such that the set of equilibria in Γ and $\bar{\Gamma}$ coincide, by the following transformation.*

Set

$$\alpha_P = -f^{\min},$$

$$\beta_e = - \min_{f^{\min} \leq f_e \leq 0} c_e(f_e),$$

$$\bar{f}_P = f_P + \alpha_P,$$

$$\bar{c}_e(\bar{f}_e) = c_e \left(\bar{f}_e - \sum_{P : e \in P} \alpha_P \right) + \beta_e,$$

$$\bar{f}^{\min} = f^{\min} + \sum_{P \in \mathcal{P}} \alpha_P,$$

$$\bar{f}_e^{\max} = f_e^{\max} + \sum_{P : e \in P} \alpha_P, \forall\, e \in E.$$

It holds that $\bar{f}$ is an equilibrium in $\bar{\Gamma}$ if and only if f is an equilibrium in Γ. The cost of player P for feasible frequencies $\bar{f}$ changes by a transformation such as the following.

$$\bar{c}_P(\bar{f}) = c_P(f) + \sum_{e \in P} \beta_e. \tag{A.1}$$

Proof. First note that β_e exists for all $e \in E$, inasmuch as $c_e(f_e)$ is continuous and thus the minimum of a compact interval exists by the Weierstrass extreme value theorem. Furthermore, α_e and β_e are nonnegative for $e \in E$ by definition.

The cost (A.1) of a line P can be verified by inserting the definitions and using $f_e = \sum_{P:e \in P} f_P = \sum_{P:e \in P} (\bar{f}_P - \alpha_P) = \bar{f}_e - \sum_{P:e \in P} \alpha_P$:

$$\bar{c}_P(\bar{f}) = \sum_{e \in P} \bar{c}_e(\bar{f}_e) = \sum_{e \in P} \left(c_e \left(\bar{f}_e - \sum_{P:e \in P} \alpha_P \right) + \beta_e \right) = c_P(f) + \sum_{e \in P} \beta_e.$$

For the proof of the equivalence of equilibria, we check the three cases that appear in the payoff function.

Part (a)

$$\bar{b}_P(\bar{f}) = M \Leftrightarrow \sum_{P \in \mathcal{P}} \bar{f}_P < \bar{f}^{\min}$$

$$\Leftrightarrow \sum_{P \in \mathcal{P}} (\bar{f}_P - \alpha_P) < \bar{f}^{\min} - \sum_{P \in \mathcal{P}} \alpha_P$$

$$\Leftrightarrow \sum_{P \in \mathcal{P}} f_P < f^{\min}$$

$$\Leftrightarrow b_P(f) = M.$$

Part (b)

$$\bar{b}_P(\bar{f}) = \bar{c}_P(\bar{f}) \Leftrightarrow \underbrace{\sum_{P \in \mathcal{P}} \bar{f}_P \geq \bar{f}^{\min}}_{(*)} \wedge \underbrace{\forall\, e \in P : \bar{f}_e \leq \bar{f}_e^{\max}}_{(**)}$$

$$\Leftrightarrow \sum_{P \in \mathcal{P}} f_P \geq f^{\min} \wedge \forall\, e \in P : f_e \leq f_e^{\max}$$

$$\Leftrightarrow b_P(f) = c_P(f).$$

$(*)\quad \sum_{P \in \mathcal{P}} \bar{f}_P \geq \bar{f}^{\min} \Leftrightarrow \sum_{P \in \mathcal{P}} \bar{f}_P - \alpha_P \geq \bar{f}^{\min} - \sum_{P \in \mathcal{P}} \alpha_P$

$$\Leftrightarrow \sum_{P \in \mathcal{P}} f_P \geq f^{\min}.$$

$(**)\ \forall\, e \in P : \bar{f}_e \leq \bar{f}_e^{\max} \Leftrightarrow \bar{f}_e - \sum_{P:e \in P} \alpha_P \leq \bar{f}_e^{\max} - \sum_{P:e \in P} \alpha_P$

$$\Leftrightarrow f_e \leq f_e^{\max}.$$

Part (c)

$$\bar{b}_P(\bar{f}) = N \Leftrightarrow \underbrace{\sum_{P \in \mathcal{P}} \bar{f}_P \geq \bar{f}^{\min}}_{(*)} \wedge \underbrace{\exists\, e \in P : \bar{f}_e > \bar{f}_e^{\max}}_{(***)}$$

$$\Leftrightarrow \sum_{P \in \mathcal{P}} f_P \leq r \wedge \exists\, e \in P : f_e > f_e^{\max}$$

$$\Leftrightarrow b_P(f) = N.$$

$$(*) \qquad\qquad \text{see } (ii)$$

$$(***)\ \exists\, e \in P : \bar{f}_e > \bar{f}_e^{\max} \Leftrightarrow \bar{f}_e - \sum_{P \in \mathcal{P}} \alpha_P > \bar{f}_e^{\max} - \sum_{P \in \mathcal{P}} \alpha_P$$

$$\Leftrightarrow f_e > f_e^{\max}.$$

Consider any two flows f_1, f_2 and the corresponding transformations $\bar{f}_1, \bar{f}_2$. As the difference of $\bar{c}_P(\bar{f})$ and $c_P(f)$ is given by a constant, and as M and N are by definition chosen sufficiently large; that is, greater than $c_P(f) + \sum_{e \in P} \beta_e$ for all feasible f, it holds:

$$\bar{b}_P(\bar{f}_1) \leq \bar{b}_P(\bar{f}_2) \quad \Leftrightarrow \quad b_P(f_1) \leq b_P(f_2).$$

Thus, the set of equilibria coincides for $\bar{\Gamma}$ and Γ. $\qquad\qquad\qquad\qquad\qquad$ □

A.2 Polyhedron for Line Planning Game

The following matrix and vector describe the polyhedron in Example 4.31.

$$
A = \left(\begin{array}{rrrrrrrrrr}
-1 & -1 & 0 & 0 & 0 & 0 & 0 & 0 & 0 & 0 \\
0 & 0 & -1 & -1 & -1 & 0 & 0 & 0 & 0 & 0 \\
0 & 0 & 0 & 0 & 0 & -1 & -1 & -1 & 0 & 0 \\
0 & 0 & 0 & 0 & 0 & 0 & 0 & 0 & -1 & -1 \\
\hline
1 & 0 & 0 & 0 & 0 & 0 & 0 & 0 & 0 & 0 \\
1 & 0 & 1 & 0 & 0 & 0 & 0 & 0 & 0 & 0 \\
1 & 0 & 0 & 0 & 0 & 0 & 0 & 0 & 0 & 0 \\
0 & 1 & 0 & 0 & 0 & 0 & 0 & 0 & 0 & 0 \\
0 & 0 & 1 & 0 & 0 & 0 & 0 & 0 & 0 & 0 \\
0 & 0 & 1 & 0 & 0 & 0 & 0 & 0 & 0 & 0 \\
0 & 1 & 0 & 0 & 0 & 0 & 0 & 0 & 0 & 0 \\
0 & 0 & 0 & 1 & 0 & 0 & 0 & 0 & 0 & 0 \\
0 & 1 & 0 & 1 & 0 & 1 & 0 & 0 & 0 & 0 \\
0 & 0 & 0 & 1 & 0 & 0 & 0 & 0 & 0 & 0 \\
0 & 0 & 0 & 0 & 1 & 0 & 0 & 0 & 0 & 0 \\
0 & 0 & 0 & 0 & 0 & 1 & 0 & 0 & 0 & 0 \\
0 & 0 & 0 & 0 & 0 & 1 & 0 & 0 & 0 & 0 \\
0 & 0 & 0 & 0 & 1 & 0 & 0 & 0 & 0 & 0 \\
0 & 0 & 0 & 0 & 0 & 0 & 1 & 0 & 0 & 0 \\
0 & 0 & 0 & 0 & 1 & 0 & 1 & 0 & 1 & 0 \\
0 & 0 & 0 & 0 & 0 & 0 & 1 & 0 & 0 & 0 \\
0 & 0 & 0 & 0 & 0 & 0 & 0 & 1 & 0 & 0 \\
0 & 0 & 0 & 0 & 0 & 0 & 0 & 0 & 1 & 0 \\
0 & 0 & 0 & 0 & 0 & 0 & 0 & 0 & 1 & 0 \\
0 & 0 & 0 & 0 & 0 & 0 & 0 & 1 & 0 & 0 \\
0 & 0 & 0 & 0 & 0 & 0 & 0 & 0 & 0 & 1 \\
0 & 0 & 0 & 0 & 0 & 0 & 0 & 1 & 0 & 1 \\
0 & 0 & 0 & 0 & 0 & 0 & 0 & 0 & 0 & 1 \\
\hline
-1 & 0 & 0 & 0 & 0 & 0 & 0 & 0 & 0 & 0 \\
0 & -1 & 0 & 0 & 0 & 0 & 0 & 0 & 0 & 0 \\
0 & 0 & -1 & 0 & 0 & 0 & 0 & 0 & 0 & 0 \\
0 & 0 & 0 & -1 & 0 & 0 & 0 & 0 & 0 & 0 \\
0 & 0 & 0 & 0 & -1 & 0 & 0 & 0 & 0 & 0 \\
0 & 0 & 0 & 0 & 0 & -1 & 0 & 0 & 0 & 0 \\
0 & 0 & 0 & 0 & 0 & 0 & -1 & 0 & 0 & 0 \\
0 & 0 & 0 & 0 & 0 & 0 & 0 & -1 & 0 & 0 \\
0 & 0 & 0 & 0 & 0 & 0 & 0 & 0 & -1 & 0 \\
0 & 0 & 0 & 0 & 0 & 0 & 0 & 0 & 0 & -1
\end{array}\right), \quad
b = \left(\begin{array}{r}
-1 \\ -1 \\ -1 \\ -1 \\ \hline
4 \\ 4 \\ 4 \\ 4 \\ 4 \\ 4 \\ 4 \\ 4 \\ 4 \\ 4 \\ 4 \\ 4 \\ 4 \\ 4 \\ 4 \\ 4 \\ 4 \\ 4 \\ 4 \\ 4 \\ 4 \\ 4 \\ 4 \\ 4 \\ \hline
0 \\ 0 \\ 0 \\ 0 \\ 0 \\ 0 \\ 0 \\ 0 \\ 0 \\ 0
\end{array}\right).
$$

References

[ABEA⁺06] E. Altman, T. Boulogne, R. El-Azouzi, T. Jiménez, and L. Wynter. A survey on networking games in telecommunications. *Computers and Operations Research*, 33:286–311, 2006.

[AD54] J. Arrow and G. Debreu. Existence of an equilibrium for a competitive economy. *Econometrica*, 22(3):265–290, 1954.

[ADK⁺04] E. Anshelevich, A. Dasgupta, J. Kleinberg, É. Tardos, T. Wexler, and T. Roughgarden. The price of stability for network design with fair cost allocation. In *Proceedings 45th IEEE Symposium on Foundations of Computer Science (FOCS)*, pages 295–304. 2004.

[AEAP02] E. Altman, R. El-Azouzi, and O. Pourtallier. Properties of equilibria in competitive routing with several users types. In *Proceedings of 41th IEEE Conference on Decision and Control*, pages 3646–3651. 2002.

[AK05] E. Altman and H. Kameda. Equilibria for multiclass routing in multi-agent networks. *Advances in Dynamic Games*, 7:343–367, 2005.

[AK07] B. Awerbuch and R. Khandekar. On cost sharing mechanisms in the network design game. In *Proceedings of the 26th annual ACM symposium on Principles of distributed computing (PODC07)*, pages 364–365, 2007.

[arr] ARRIVAL: *Algorithms for robust and online railway optimization: improving the validity and reliability of large scale systems.* Funded by the "Future and Emerging Technologies" Unit of the EC within the 6th Framework Programme of the European Commission, http://arrival.cti.gr/.

[AT02] A. Archer and É. Tardos. Frugal path mechanisms. In *Proceedings of the 13th Annual ACM-SIAM Symposium on Discrete Algorithms*, pages 991–999. ACM Press, New York, 2002.

[BCKV06] R. Beier, A. Czumaj, P. Krysta, and B. Vöcking. Computing equilibria for a service provider game with (im)perfect information. *ACM Transactions on Algorithms*, 2(4):679–706, 2006.

[BDHS02] R. Baron, J. Durieu, H. Haller, and P. Solal. Control costs and potential functions for spatial games. *International Journal of Game Theory*, 31:541–561, 2002.

[BE05] U. Brandes and T. Erlebach, editors. *Network Analysis*, volume 3418 of *Lecture Notes in Computer Science*. Springer, Berlin, 2005.

[BGP04a]   R. Borndörfer, M. Grötschel, and M. E. Pfetsch. Models for line planning in public transport. Technical Report 04-10, ZIP Berlin, 2004.

[BGP04b]   R. Borndörfer, M. Grötschel, and M.E. Pfetsch. A path-based model for line planning in public transport. Technical Report 05-18, ZIP Berlin, 2004.

[Bil98]   J.-M. Bilbao. Values and potential of games with cooperation structure. *International Journal of Game Theory*, 27:131–145, 1998.

[Bir76]   C.G. Bird. On cost allocation for a spanning tree: A game theoretic approach. *Networks*, 6(4):335–350, 1976.

[BKZ96]   M.R. Bussieck, P. Kreuzer, and U.T. Zimmermann. Optimal lines for railway systems. *European Journal of Operational Research*, 96(1):54–63, 1996.

[BLPGVR04]   Y. Bramoullé, D. López-Pintado, S. Goyal, and F. Vega-Redondo. Network formation and anti-coordination games. *International Journal of Game Theory*, 33(1):1–19, 2004.

[BMI06]   C. Busch and M. Magdon-Ismail. Atomic routing games on maximum congestion. In *Algorithmic Aspects in Information and Management*, volume 4041 of *Lecture Notes in Computer Science*, pages 79–91. Springer, Berlin, 2006.

[BS79]   M. Bazaraa and C.M. Shetty. *Nonlinear Programming*. Wiley, New York, 1979.

[BS91]   I.N. Bronstein and K.A. Semendjajew. *Taschenbuch der Mathematik*. Nauka, Teubner, Moskau, Leipzig, 1991.

[BS02]   T. Başar and R. Srikant. A stackelberg network game with a large number of followers. *Journal of Optimization Theory and Applications*, 155(3):479–490, 2002.

[BS06]   N. Baumann and M. Skutella. Solving evacuation problems efficiently–earliest arrival flows with multiple sources. In *Proceedings of the 47th Annual IEEE Symposium on Foundations of Computer Science (FOCS06)*, pages 399–410, 2006.

[CC01]   R. Cominetti and J. Correa. Common-lines and passenger assignment in congested transit networks. *Transportation Science*, 35(3):250–267, 2001.

[CCLE+06]   C.Chekuri, J. Chuzhoy, L. Lewin-Eytan, J. Naor, and A. Orda. Non-cooperative multicast and facility location games. In *Proceedings of the 7th ACM conference on Electronic commerce (EC06)*, pages 72–81, 2006.

[CCSM06]   R. Cominetti, J.R. Correa, and N.E. Stier-Moses. Network games with atomic players. In *Automata, Languages and Programming*, volume 4051 of *Lecture Notes in Computer Science*, pages 525–536. Springer, Berlin, 2006.

[CDR03]   R. Cole, Y. Dodis, and T. Roughgarden. Pricing network edges for heterogeneous selfish users. In *Proceedings of the 35th Annual ACM Symposium on Theory of Computing (STOC)*, pages 521–530. ACM Press, New York, 2003.

[CGY99]   G.Y. Chen, C.J. Goh, and X.Q. Yang. Vector network equilibrium problems and nonlinear scalarization methods. *Mathematical Methods of Operations Research*, 49:239–253, 1999.

[CKV02]   A. Czumaj, P. Krysta, and B. Vöcking. Selfish traffic allocation for
           server farms. In *Proceedings of the 34th ACM Symposium on Theory
           of Computing (STOC)*, pages 287–296. 2002.

[CLPU04]   L. Castelli, G. Longo, R. Pesenti, and W. Ukovich. Two-player nonco-
           operative games over a freight transportation network. *Transportation
           Science*, 38(2):149–159, 2004.

[CP82]   D. Chan and J.-S. Pang. The generalized quasi-variational inequality
           problem. *Mathematics of Operations Research*, 7(2):211–220, 1982.

[CR06]   H.L. Chen and T. Roughgarden. Network design with weighted play-
           ers. In *Proceedings of the 18th Annual ACM Symposium on Paral-
           lelism in Algorithms and Architectures (SPAA06)*, pages 29–38. 2006.

[CRV08]   H.-L. Chen, T. Roughgarden, and G. Valiant. Designing networks
           with good equilibria. In *Proceedings of the 19th annual ACM-SIAM
           symposium on Discrete algorithms (SODA '08)*, pages 854–863, 2008.

[CSS04a]   J.R. Correa, A.S. Schulz, and N.E. Stier Moses. Computational com-
           plexity, fairness, and the price of anarchy of the maximal latency
           problem. In *Integer Programming and Combinatorial Optimization,
           10th International IPCO Conference*, pages 59–73. 2004.

[CSS04b]   J.R. Correa, A.S. Schulz, and N.E. Stier Moses. Selfish routing in
           capacitated networks. *Mathematics of Operations Research*, 29(4):
           961–976, 2004.

[CV07]   A. Czumaj and B. Vöcking. Tight bounds for worst case equilibria.
           ACM Transactions on Algorithms, 3(1):4, 2007.

[Deb52]   G. Debreu. A social equilibrium existence theorem. *Proceedings of
           the National Academy of Sciences*, 38(10):886–893, 1952.

[DGK+05]   N. Devanur, N. Garg, R. Khandekar, V. Pandit, A. Saberi, and
           V.V. Vazirani. Price of anarchy, locality gap, and a network ser-
           vice provider game. In *Proceedings of the First Workshop on Internet
           and Network Economics (WINE)*, pages 1046–1055. 2005.

[DH81]   D.Granot and G. Huberman. Minimum cost spanning tree games.
           Mathematical Programming, 21(1):1–18, 1981.

[DR02]   T.S.H. Driessen and T. Radzika. A weighted pseudo-potential ap-
           proach to values for tu-games. *International Transactions in Opera-
           tional Research*, 9:303–320, 2002.

[EBKS07]   E. Even-Bar, M. Kearns, and S. Suri. A network formation game
           for bipartite exchange economies. In *Proceedings of the 18th an-
           nual ACM-SIAM symposium on Discrete algorithms (SODA)*, pages
           697–706, 2007.

[Ehr05]   M. Ehrgott. *Multicriteria Optimization*. Springer, Berlin, 2005.

[ESS04]   E. Elkind, A. Sahai, and K. Steiglitz. Frugality in path auctions. In
           *Proceedings of the 15th Annual ACM-SIAM Symposium on Discrete
           Algorithms*, pages 701–709. ACM Press, New York, 2004.

[FKK+02]   D. Fotakis, S.C. Kontogiannis, E. Koutsoupias, M. Mavronicolas, and
           P.G. Spirakis. The structure and complexity of Nash equilibria for
           a selfish routing game. In *Proceedings of the 29th International Col-
           loquium on Automata, Languages and Programming (ICALP)*, pages
           123–134. 2002.

[Fle04]   L. Fleischer. Linear tolls suffice: New bounds and algorithms for tolls
           in single source networks. In *Automata, Languages and Programming:*

31st International Colloquium, (ICALP), pages 544–554. Springer, New York, 2004.

[FLM+03] A. Fabrikant, A. Luthra, E. Maneva, C.H. Papadimitriou, and S. Shenker. On a network creation game. In *Proceedings of the 22nd Annual Symposium on Principles of Distributed Computing*, pages 347–351. 2003.

[FMBT97] G. Facchini, F.v. Megen, P. Borm, and S. Tijs. Congestion models and weighted Bayesian potential games. *Theory and Decision*, 42:193–206, 1997.

[FP05] M. Fukushima and J.-S. Pang. Quasi-variational inequalities, generalized Nash equilibria, and multi-leader-follower games. *Computational Management Science*, 2(1):21–56, 2005.

[FP06] F. Facchinei and J.-S. Pang. Exact penalty functions for generalized Nash problems. In G. Di Pillo and M. Roma, editors, *Large-scale Nonlinear Optimization*, pages 115–126. Springer, New York, 2006.

[FT91] D. Fudenberg and J. Tirole. *Game Theory*. MIT Press, Cambridge, MA, 1991.

[FV04] S. Fischer and B. Vöcking. On the evolution of selfish routing. In *Proceedings of the 12th Annual European Symposium on Algorithms (ESA)*, pages 323–334. 2004.

[FV05] S. Fischer and B. Vöcking. On the structure and complexity of worst-case equilibria. In *Proceedings of the 1st Workshop on Internet and Network Economics (WINE)*, pages 151–160. 2005.

[GC05] D. Grosu and A.T. Chronopoulos. Noncooperative load balancing in distributed systems. *Journal of Parallel and Distributed Computing*, 65(9):1022–1034, 2005.

[GGAM+03] D. Gómez, E. González-Arangüena, C. Manuel, G. Owen, M. del Pozo, and J. Tejada. Centrality and power in social networks: A game theoretic approach. *Mathematical Social Sciences*, 46:27–54, 2003.

[GKR03] A. Gupta, A. Kumar, and T. Roughgarden. Simpler and better approximation algorithms for network design. In *Proceedings of the 35th Annual ACM Symposium on Theory of Computing*, pages 365–372. 2003.

[GO82] B. Grofman and G. Owen. A game theoretic approach to measuring centrality in social networks. *Social Networks*, 4(3):213–224, 1982.

[GS04] M.X. Goemans and M. Skutella. Cooperative facility location games. *Journal of Algorithms*, 50(2):194–214, 2004.

[GVR05] S. Goyal and F. Vega-Redondo. Network formation and social coordination. *Games and Economic Behavior*, 50:178–207, 2005.

[GY99] C.J. Goh and X.Q. Yang. Vector equilibrium problem and vector optimization. *European Journal of Operational Research*, 116:615–628, 1999.

[Har91] P.T. Harker. Generalized Nash games and quasi-variational inequalities. *European Journal of Operational Research*, 54:81–94, 1991.

[HK00] H.W. Hamacher and K. Klamroth. *Linear and Network Optimization*. Vieweg, Braunschweig, Wiesbaden, 2000.

[HMC89] S. Hart and A. Mas-Colell. Potential, value, and consistency. *Econometrica*, 57:589–614, 1989.

[HS88]   J.C. Harsanyi and R. Selten. *A General Theory of Equilibrium Selection in Games*. MIT Press, Cambridge, MA, 1988.
[HSK02]   D. Helbing, M. Schönhof, and D. Kern. Volatile decision dynamics: Experiments, stochastic description, intermittency control and traffic optimization. *New Journal of Physics*, 4(33):1–16, 2002.
[HT01]   H.W. Hamacher and S.A. Tjandra. Mathematical modelling of evacuation problems: A state of art. Berichte des Fraunhofer ITWM 24, Fraunhofer ITWM, Kaiserslautern, 2001.
[HTW05]   A. Hayrapetyan, É. Tardos, and T. Wexler. A network pricing game for selfish traffic. In *Proceedings of the 24th Annual ACM SIGACT-SIGOPS Symposium on Principles of Distributed Computing (PODC)*, pages 284–289. ACM Press, New York, 2005.
[Jac05]   M.O. Jackson. Allocation rules for network games. *Games and Economic Behavior*, 51:128–154, 2005.
[JMSSM05] O. Jahn, R.H. Möhring, A.S. Schulz, and N.E. Stier-Moses. System-optimal routing of traffic flows with user constraints in networks with congestion. *Operations Research*, 53(4):600–616, 2005.
[JS08]   B.H. Junker and F. Schreiber, editors. *Analysis of Biological Networks*. Wiley & Sons, New York, 2008.
[JT04]   R. Johari and J.N. Tsitsiklis. Efficiency loss in a network resource allocation game. *Mathematics of Operations Research*, 29(3): 407–435, 2004.
[Kel97]   F. Kelly. Charging and rate control for elastic traffic. *European Transactions on Telecommunications*, 8:33–37, 1997.
[KK03]   G. Karakostas and S.G. Kolliopoulos. Selfish routing in the presence of side constraints. Technical Report CAS-03-13-GK, Department of Computing and Software, McMaster University, 2003.
[KLO97]   Y.A. Korilis, A.A. Lazar, and A. Orda. Achieving network optima using stackelberg routing strategies. *IEEE/ACM Transactions on Networking*, 5(1):161–173, 1997.
[KLS05]   J. Könemann, S. Leonardi, and G. Schäfer. A group-strategyproof mechanism for steiner forests. In *Proceedings of the 16th annual ACM-SIAM symposium on Discrete algorithms (SODA05)*, pages 612–619, 2005.
[KP99]   E. Koutsoupias and C. Papadimitriou. Worst-case equilibria. In *Proceedings of the 16th Annual Symposium on Theoretical Aspects of Computer Science (STACS'99)*, pages 404–413. Springer, New York, 1999.
[Kra05]   W. Krabs. *Spieltheorie*. Teubner, Wiesbaden, 2005.
[KS06]   A.C. Kaporis and P.G. Spirakis. The price of optimum in stackelberg games on arbitrary single commodity networks and latency functions. In *Proceedings of the 18th annual ACM symposium on Parallelism in algorithms and architectures (SPAA06)*, pages 19–28, 2006.
[Kuk99]   N.S. Kukushkin. Potential games: A purely ordinal approach. *Economics Letters*, 64:279–283, 1999.
[Kuk02]   N.S. Kukushkin. Perfect information and potential games. *Games and Economic Behavior*, 38:306–317, 2002.
[LMMR04]   T. Lücking, M. Mavronicolas, B. Monien, and M. Rode. A new model for selfish routing. In *Proceedings of the 21st Annual Symposium on Theoretical Aspects of Computer Science*, pages 547–558. 2004.

[Mal07]   L. Mallozzi. Noncooperative facility location games. *Operations Research Letters*, 35(2):151–154, 2007.

[Mil96]   I. Milchtaich. Congestion games with player-specific payoff functions. *Games and Economic Behavior*, 13:111–124, 1996.

[Mon07]   D. Monderer. Multipotential games. In *Proceedings of 20th International Joint Conference on Artificial Intelligence (IJCAI)*, pages 1422–1427. 2007.

[MS96]   D. Monderer and L. S. Shapley. Potential games. *Games and Economic Behavior*, 14:124–143, 1996.

[MTV00]   L. Mallozzi, S. Tijs, and M. Voorneveld. Infinite hierarchical potential games. *Journal of Optimization Theory and Applications*, 107(2):287–296, 2000.

[MU05]   S. Morris and T. Ui. Generalized potentials and robust sets of equilibria. *Journal of Economic Theory*, 124:45–78, 2005.

[Nag00]   A. Nagurney. A multiclass, multicriteria traffic network equilibrium model. *Mathematical and Computer Modelling*, 32:393–411, 2000.

[Nas50]   J. Nash. Equilibrium points in n-person games. *Proceedings of the National Academy of Sciences*, 36:48–49, 1950.

[Nis99]   N. Nisan. Algorithms for selfish agents. In *Proceedings of the 16th Annual Symposium on Theoretical Aspects of Computer Science (STACS99)*, volume 1563 of *Lecture Notes in Computer Science*, pages 1–15. Springer, Berlin, 1999.

[NIS05]   E. Nikolova N. Immorlica, D. Karger and R. Sami. First-price path auctions. In *Proceedings of 6th ACM Conference on Electronic Commerce (EC05)*, pages 203–212, 2005.

[NR01]   N. Nisan and A. Ronen. Algorithmic mechanism design. *Games and Economic Behavior*, 35:166–196, 2001.

[NT98]   H. Norde and S. Tijs. Determinateness of strategic games with a potential. *Mathematical Methods of Operations Research*, 46:377–385, 1998.

[Owe95]   G. Owen. *Game Theory*. Academic Press, San Diego, third edition, 1995.

[Pap01]   C.H. Papadimitriou. Algorithms, games, and the internet. In *Proceedings of the 33th ACM Symposium on Theory of Computing (STOC)*, pages 749–753. 2001.

[Per04]   G. Perakis. The price of anarchy when costs are non-separable and asymetric. In *Integer Programming and Combinatorial Optimization, 10th International IPCO Conference*, pages 46–58. Springer, New York, 2004.

[PPT07]   F. Patrone, L. Pusillo, and S. Tijs. Multicriteria games and potentials. *TOP*, 15(1):138–145, 2007.

[PSS]   J. Puerto, A. Schöbel, and S. Schwarze. Games on polyhedra. Working paper.

[PSS06]   J. Puerto, A. Schöbel, and S. Schwarze. A class of infinite potential games. Technical Report 2006-15, Institut für Numerische und Angewandte Mathematik, Universität Göttingen, 2006. Preprint Reihe.

[PSS08]   J. Puerto, A. Schöbel, and S. Schwarze. The path player game: A network game from the point of view of the network providers. *Mathematical Methods of Operations Research*, 2008. to appear.

[Ros65]   J.B. Rosen. Existence and uniqueness of equilibrium points for concave n-person games. *Econometrica*, 33(3):520–534, 1965.

[Ros73]   R.W. Rosenthal. A class of games possessing pure-strategy Nash equilibria. *International Journal of Game Theory*, 2:65–67, 1973.

[Rou03]   T. Roughgarden. The price of anarchy is independent of the network topology. *Journal of Computer and System Sciences*, 67(2):341–364, 2003.

[Rou04]   T. Roughgarden. Stackelberg scheduling strategies. *SIAM Journal on Computing*, 33(2):332–350, 2004.

[Rou05a]   T. Roughgarden. *Selfish Routing and the Price of Anarchy*. MIT Press, Cambridge, MA, 2005.

[Rou05b]   T. Roughgarden. Selfish routing with atomic players. In *Proceedings of the 16th annual ACM-SIAM symposium on Discrete algorithms (SODA05)*, pages 1184–1185, 2005.

[RT02]   T. Roughgarden and É. Tardos. How bad is selfish routing? *Journal of the ACM*, 49(2):236–259, 2002.

[San01]   W.S. Sandholm. Potential games with continuous player sets. *Journal of Economic Theory*, 97:81–108, 2001.

[Sch05]   S. Scholl. *Customer-oriented line planning*. PhD thesis, Technische Universität Kaiserslautern, 2005.

[SDNT00]   M. Slikker, B. Dutta, A.v.d. Nouweland, and S. Tijs. Potential maximizers and network formation. *Mathematical Social Sciences*, 39: 55–70, 2000.

[SK95]   D. Skorin-Kapov. On the core of the minimum cost Steiner tree game in networks. *Annals of Operations Research*, 57:233–249, 1995.

[Sli01]   M. Slikker. Coalition formation and potential games. *Games and Economic Behavior*, 37:436–448, 2001.

[SS02]   M. Schreckenberg and S.D. Sharma, editors. *Pedestrian and Evacuation Dynamics*. Springer, Berlin, 2002.

[SS05]   A. Schöbel and S. Scholl. Line planning with minimal transfers. In *Proceedings of 5th Workshop on Algorithmic Methods and Models for Optimization of Railways (ATMOS)*. 2005.

[SS06a]   A. Schöbel and S. Schwarze. Dominance and equilibria in the path player game. In *Operations Research Proceedings 2005*, pages 489–494. 2006.

[SS06b]   A. Schöbel and S. Schwarze. A game-theoretic approach to line planning. In *ATMOS 2006-6th Workshop on Algorithmic Methods and Models for Optimization of Railways*, number 06002 in Dagstuhl Seminar Proceedings, 2006. http://drops.dagstuhl.de/opus/volltexte/2006/688.

[STZ04]   S. Suri, C.D. Tóth, and Y. Zhou. Selfish load balancing and atomic congestion games. In *Proceedings of the 16th annual ACM symposium on Parallelism in algorithms and architectures (SPAA04)*, pages 188–195. 2004.

[Swa07]   C. Swamy. The effectiveness of stackelberg strategies and tolls for network congestion games. In *Proceedings of the 18th annual ACM-SIAM symposium on Discrete algorithms (SODA07)*, pages 1133–1142, 2007.

[Ui00]   T. Ui. A Shapley value representation of potential games. *Games and Economic Behavior*, 31:121–135, 2000.

[Ui01]   T. Ui.   Robust equilibria of potential games.   *Econometrica*, 69(5):1373–1380, 2001.

[Vet02]   A. Vetta. Nash equilibria in competitive societies, with applications to facility location, traffic routing and auctions. In *Proceedings of the 43rd Symposium on Foundations of Computer Science (FOCS)*, pages 416–425. 2002.

[VN97]   M. Voorneveld and H. Norde. A characterization of ordinal potential games. *Games and Economic Behavior*, 19:235–242, 1997.

[Voo97]   M. Voorneveld. Equilibria and approximate equilibria in infinite potential games. *Economics Letters*, 56:163–169, 1997.

[Voo00]   M. Voorneveld. Best-response potential games. *Economics Letters*, 66:289–295, 2000.

[War52]   J.G. Wardrop. Some theoretical aspects of road traffic research. *Proceedings of the Institute of Civil Engineers, Pt. II*, 1:325–378, 1952.

[Yan07]   Q. Yan. On the price of truthfulness in path auctions. In *Internet and Network Economics*, volume 4858 of *Lecture Notes in Computer Science*, pages 584–589. Springer, Berlin, 2007.

Index